Fenland: its ancient past and
uncertain future

Fenland: its ancient past and uncertain future

SIR HARRY GODWIN FRS

EMERITUS PROFESSOR OF BOTANY

CAMBRIDGE UNIVERSITY

CAMBRIDGE UNIVERSITY PRESS

CAMBRIDGE

LONDON · NEW YORK · MELBOURNE

CAMBRIDGE UNIVERSITY PRESS
Cambridge, New York, Melbourne, Madrid, Cape Town, Singapore, São Paulo, Delhi

Cambridge University Press
The Edinburgh Building, Cambridge CB2 8RU, UK

Published in the United States of America by Cambridge University Press, New York

www.cambridge.org
Information on this title: www.cambridge.org/9780521103398

First published 1978
This digitally printed version 2009

A catalogue record for this publication is available from the British Library

Library of Congress Cataloguing in Publication data
Godwin, Sir Harry, 1901–
Fenland: its ancient past and uncertain future

Bibliography: p. 186   Includes index
1. Geology – England – Fens   2. Botany – England – Fens – Ecology
3. Fens, England   I. Title

QE262. F3G63   500.9′426   77-8824

ISBN 978-0-521-21768-2 hardback
ISBN 978-0-521-10339-8 paperback

Contents

Acknowledgments   *page* vii

1   Introduction   1
2   Ecological background   9
3   Pollen analysis   21
4   Bog oaks and buried forests   33
5   Flandrian deposits and the Fenland Research Committee   43
6   Shippea Hill and the natural bed of the River Little Ouse   50
7   The Lower Peat and the Fen Clay   60
8   The Upper Peat: hoards and trackways   68
9   Iron Age hiatus, roddons and Romans   79
10 Extinct meres and shell-marl   91
11 Conspectus and historical framework   102
12 Peat and its winning   111
13 The loss of the peat: shrinkage and wastage   124
14 Fenland drainage   134
15 Ancient crops, natural and cultivated   145
16 Lost and vanishing species: conservation   164

References   186
Index   188

Acknowledgments

The number of friends who have assisted me through the years on one kind of Fenland project or another is legion, particularly colleagues and associates in the former Fenland Research Committee and among a sequence of research students, research assistants, laboratory technicians and staff colleagues, who enthusiastically shared in a variety of projects, some deep and persisting, others no more than a hard day's field-work with spade, peat-auger and level, or the determination of an obscure plant fragment or other fossil. They will forgive me if I do not repeat the individual thanks to them that I have already expressed elsewhere: my gratitude to them is nevertheless still vivid.

More specifically in preparation of this book I owe thanks to Mr G. Pearson, Chief Engineer of the Great Ouse River Division of the Anglian Water Authority for bringing me up to date with information on the most recently completed major Fenland drainage works, and to Mr L. F. Fillenham, Chief Engineer to the Middle Level Commissioners, for his similar rôle in informing me of his research on, and restoration of the famous Holme Fen post.

Again specifically I am happy to thank those who kindly provided me with Fenland photographs, Mr W. H. Palmer, Dr M. C. F. Proctor, Dr C. Forbes, Mr P. Sell, Professor G. E. Briggs, Professor J. K. S. St Joseph, Director of the University Committee for Aerial Photography, Professor H. C. Darby and Dr B. W. Sparks: I owe a great deal also to the precision and skill of Mr F. T. N. Elborn and Miss Sylvia Bishop in copying and reducing to monochrome so many of my own pictures.

Finally, and most comprehensively I have to thank my wife for participation and support through more than fifty years of Fenland investigations. From the start we saw that pollen analysis might provide the much-needed chronological scale in the study of Fenland development and pre-history. Indeed by the time that the Fenland Research Committee was founded she had already demonstrated its feasibility, and together we thereafter co-operated in most of its activities, as in similar studies elsewhere in Britain. Large as this contribution has been it is but a small part of my continuing debt to her, now gratefully if inadequately acknowledged.

I

Introduction

The Fenland of East Anglia is the vast shallow basin, several hundred
square miles in extent, stretching from Lincoln in the north to Cambridge in
the south, and from Peterborough and Huntingdon in the west to Boston
and King's Lynn close to the coast of the Wash. The whole structure and
economy of the area has been determined by the superfluity of water
brought into it on the one hand by inundations of sea-water, particularly
during times of a rise of sea-level relative to the land, and on the other by
fresh-water from an extremely large catchment area of surrounding upland
that is concentrated within it by the flow of such considerable rivers as the
Witham, Welland, Nene, Great Ouse, Little Ouse, Lark and Nar. Defoe in
1724 had summed up the situation perfectly: 'in a word, all the water of the
middle part of England which does not run into the Thames or Trent comes
down into these fens'. The Fen basin could fairly be said to be brimful of
deposits laid down in water, silts and clays on the seaward side and black
fresh-water peat behind them. It is this infilling which has induced that
outstanding, and to some, entrancing quality of the Fenland landscape, an
interminable flatness relieved only here and there by the gentle emergence
of low islands of gravel where the fen floor pokes upward above the general
water-level. To travel by train north from Ely is to be reminded of the vast
Hungarian plain, likewise devoid of trees save small clumps planted round
the isolated farms, and equally without hedges: both convey feelings of
vastness and remoteness, although the dominant factor of the one landscape
is drought and the other wetness, and a closer look shews the Fenlands to be
carrying far more diverse crops and, in response to the prevalence of water,
to be regularly and closely intersected by drainage dykes. In both the
flatness provides that feature of a display of the vast hemisphere of sky
impossible to match elsewhere in Britain and the key to the affection the
landscape generates in the hearts of resident fenmen. As one, unexpectedly
communicative, explained to me: 'any fool can appreciate mountain scenery
but it takes a man of discernment to appreciate the Fens'. However the car,
the tractor and the roads they require, together with accelerated drainage
and wastage of the peat, are reducing both the scale and isolation of the

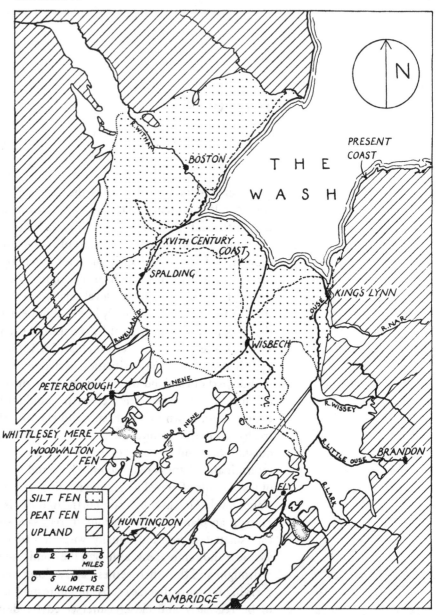

Fig. 1. Map of the Fenland basin showing in broadest outline the distribution of the main areas of peat and silt and the islands of basal clay and gravel. The former main estuary is indicated by the silt tapering south-eastwards from Wisbech across the line of the Bedford Levels towards Littleport.

Fens, so that it seems a far cry to the quite recent time when the fen waterways were a major means of transportation and when in winter it was easier to travel from Cambridge to Ely by boat than by road, and when for the whole winter the wife of a fenman working deep in the heart of the black fens would be regularly cut off with her children, by the impassibility of the peaty fen droves, from all contact with the villages and towns of the uplands – a state of affairs persisting into the 1920s. All the same, it is still possible to

Plate *1*. View of Ely from Stuntney Old Hall across the neck of ploughed peat-fen in which was found the Late Bronze Age founder's hoard.

see the reassuring command of Ely cathedral visible on its island from miles across the Fenland and to realise that St Guthlac may indeed have built his refuge at Crowland in a 'hideous fen of a huge bigness' despite 'such apparitions of devils as were so frequently seen there'.

The powerful visual appeal of this singular landscape, with its capacity to arouse strong emotional responses in us, is paralleled by the appeal it makes to the scientist, the archaeologist and the historian through the wealth of opportunity it presents for exploration of fresh evidence in their respective fields. This became most strongly apparent about a century ago, especially through the publication of the very comprehensive volume *The Fenland, Past and Present* by S. H. Miller and S. B. J. Skertchly (1878). The second of these authors had published in the previous year his Memoir of the Geological Survey, *The Geology of the Fenland*, a work of remarkable virtuosity and foresight that, by venturing to go counter to fashionable geological tenets of the time, came so close to the true interpretation of the geology of the Fenland that in essentials it has never been successfully challenged. It has moreover the great virtue that it records objectively a great deal of evidence no longer accessible to us. Miller has been less fortunate in that the great reorganisation of our knowledge of British prehistoric archaeology was to come well after the turn of the century and we find overmuch attribution by him and his contemporaries to conjectural Roman activity, and that their documentary research of later periods was far less stringently based than is now acceptable. All the same Miller provides,

Plate 2. Typical view of ploughed peat-fen at Ramsey Hollow near Woodwalton. In the foreground fruiting heads of the giant reed, *Phragmites communis*: the level plough land is broken only by the lines of drainage ditches and the trees indicate local elevations.

in the joint volume, valuable accounts of the contemporary fauna and flora, of industries such as peat-winning, of the operation of wild fowl decoys and of the past prevalence of particular diseases such as malaria and phthisis.

After the appearance of these great pioneer works no general reappraisal of the history of the Fenland was made for a very long time although scientific technology was increasingly applied to the building of railways, to drainage and water-transport and to the means of cultivation. Continuing discoveries made during these operations were separately recorded, but it was not until the present century was well into its stride that acceptance of a new scientific principle provided the mechanism allowing re-assessment and unification of the whole historical position. This came with the development of the subject of ecology and its growing acceptance from the 1920s onwards, and more especially the recognition of the validity of the ecological approach to all problems of the environment, that now in the 1970s we all take for granted. In understanding how the Fenlands came into being and what their natural condition was like, the ecologist now had advantages over the classical geologists who had hitherto been the outstanding interpreters of the Fens. Although the geologist of that period had based his science upon the principle that the past history of the earth is to be explained in terms of natural processes still at work, he was naturally experienced almost wholly with geological events and periods operating over very long periods of time, often tens of thousands of years and with considerable spatial dimensions, tens or hundreds of feet (or metres) in thickness of beds or in elevation or depression. Phenomena of a few seasons only, the local shift of a river channel, a single tidal cycle or the lifetime of one organism, concerned him only in a general way and dimensions of a few inches (centimetres) rise or fall in water-level or mud-accumulation concerned him only incidentally. On the other hand the field-ecologist finds such 'minor' changes and dimensions crucial to the lives of the organisms with which he deals and vital to the control of both plant and animal communities, so that he has perforce to become an investigator of geological events on a small and very local scale. With the rise of ecology, his experience came to fill a gap in the scale of geological expertise, an expertise to which otherwise he was nevertheless constantly in debt.

The relevance of these considerations lies of course in the comparatively minor and recent character of the geological events that brought the Fenlands into being and determined their nature: in fact we now appreciate that the bulk of Fenland deposits have been laid down within so short a time as the last six thousand years, and many of them reflect processes and changes of familiar ecological patterns. If it is to be successful this book must show how effective the marriage of the ecological and geological disciplines has been in making out the story of the Fens.

It is impossible to comprehend this story without the realisation that the Fenland deposits of the last six thousand years owe their origin primarily to a major rise in ocean level, and that their variations are largely due to subsequent lesser movements of relative height of land and sea. We may summarise the consequences of this over-riding control in the following way.

During periods of marine invasion the tidal water caused backing-up of the river water behind it so producing waterlogging over the whole hinterland, and there the landscape was one of continuous fen vegetation whose remains stay largely undecayed to form peat. This is the key to the present separation of the Fenland into the wide seaward belt of silt formed in estuarine conditions, and the black peat fens between it and the Fen margin. It supplies also the key to understanding the stratigraphy of the Fen deposits and so unravelling the geological history of the Fens: in periods of marine transgression the silts and clays will have overlaid previously deposited peats, and during emergences freshwater peat will have extended seaward over the silts and clays. Some oscillation in former relative land and sea level has therefore produced an alternate wedging-out of silts and clays on the one hand and of peats on the other (Fig. 2). Near the Fenland margin, beyond reach of even the furthest marine incursion, there is uninterrupted peat to a depth even now in places of 12 or 14 ft (3 or 4 m), further seaward the peat is split into two by an increasing thickness of the estuarine 'buttery clay', and still nearer the Wash the silts and clays are only broken by thin

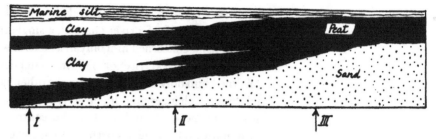

Fig. 2. Diagram to show deposits of the north-west German coast revealed by borings at the three sites indicated. The alternation of marine clays and silts with fresh-water peat is quite similar to that in the Fenland basin on the opposite shore of the North Sea.

layers of peat. We shall see how knowledge of this can be systematised and correlated with biological changes and phases of human occupation over the last seven or eight thousand years.

It was by a fortunate series of developments that I became so deeply involved in the decipherment of the Fenland story. A very early and continuing involvement with ecological teaching and research was, in my case, conveniently developed in the residual Fenland of East Anglia and especially at Wicken Fen, so that a series of observations and publications on fen ecology by myself or by students whom I directed has continued through from 1923 until 1974, and these have also extended to some other parts of Britain. The following chapter indicates how basic these continuing studies of natural plant communities and their component plants have proved to be in the investigation of Fenland history. The second major influence concerned my later, but equally deep and persistent interest in the technique of 'pollen analysis', rightly regarded as 'palaeoecology'. As is described in Chapter 3, this powerful research tool was introduced into the Godwin ménage by my wife in the early 1930s. It was initially applied to samples taken from deep peat deposits such as were naturally available in the Fens, and as it became evident that its findings were going to be closely bound up with knowledge of fen and bog vegetation it was progressively harder to resist the opportunity to involve myself with it, particularly because the fieldwork of excavation and boring required for stratigraphic information sampling was often very demanding physically. Before long in fact 'H. and M. E. Godwin' constituted an active pollen-analytic team, responsible for several early publications, and the basis of a great deal of subsequent scientific activity centred on the Cambridge Botany School. One should add a concern with, and affection for geology active already when, unusually, it formed one of my entrance scholarship subjects as long ago as 1918; it was a trend happily fostered by a series of very distinguished senior geologists who also regarded recent Fenland deposits as meriting fuller attention than they had hitherto had.

It is not hard, in retrospect, to see how this combination of scientific interests led naturally, almost inevitably, to the sequence of Fenland investigations that are covered in Chapters 2 to 4, all of them based on drawing jointly upon plant ecology, pollen analysis and stratigraphy. Meanwhile a remarkable revitalisation was becoming manifest in British archaeology where professionals and amateurs were becoming acutely aware of the great advantages that could follow integration with the natural sciences. We were now in the fortunate position of participation in the enthusiastic group of specialists that constituted the Fenland Research Committee, whose origin in 1932, and whose nature, are described in Chapter 5. In the few short years of its existence it took advantage of fortuitous discoveries and made opportunities for research projects in the Fenlands that were startlingly successful examples of scientific team work in the field and laboratory. We brought the combined resources of archaeological scholarship, and of the biological, geographical and geological sciences jointly to bear, and with enthusiasm addressed ourselves to unravelling the complex story of the Fenland. To a considerable extent the body of this book is a narrative of the progress thus made, and continued more sporadically by some of us after dissolution of the Committee itself. If the account has a 'who dun it?' quality this is entirely proper: that is the quality of all active scientific research, which inevitably looks for causes and mechanisms, and it certainly marked all the meetings, field investigations and publications of the committee.

The history of our quest to establish a solid factual basis for our ideas on the structure, evolution and cultures of the Fenland basin occupies the whole of Chapters 5 to 11. Despite the importance and coherence of this theme it represents only one phase in the history of modern Fenland research. In the years since those when the Fenland Research Committee were active a good deal of further progress has been made. It has become plain that we much underestimated the diversity of bog- and peat-types that were formerly present in the Fens. This factor, however, is one on which ancient history, past economy, modern usages and the natural occurrences of plant and animal life all depend so that it is important to record what evidence we have for it. Likewise knowledge of the nature of peat shrinkage and loss have been greatly extended by observations during their continued operation: their consequences for the whole Fenland scene and livelihood are so severe that they cannot be disregarded. I have accordingly included accounts of three main categories of changes thus induced. Firstly an outline of the principles of development of the man-made drainage system, both as causing peat loss and as reacting to it. Secondly, an account of a few highly characteristic Fenland crops, now vanished or rare, and thirdly a brief consideration of the extensive losses and threats of losses that these vast and rapid changes have brought about in the natural fauna and flora of the region.

It has been incidental that the continuing series of Fenland investigations through fifty years or so has taken one into tracts of Fen country often extremely remote and unfamiliar to outsiders. Our visits were of necessity made in all kinds of weather depending upon the chance of some discovery or other, a temporary section or the needs of a fixed programme of observations. To work in the wet clays and soft peats 20 ft below sea-level in some sluice excavation in misty rain, gives one what can only be called an 'intimate' familiarity with fen conditions. Likewise if one undertakes hourly records of water-level from dawn to dusk through hot summer days in lush vegetation and still air infested with hungry mosquitos. Particular enquiries take one to unfamiliar droves and drainage channels, and one learns to appreciate the special qualities and skills of the fen folk. Above all one generates an appreciation of and affection for the vast hemispherical skyscapes, and one develops a great deal of interest in the evidences of ways of life, customs, architecture and crops that have recently vanished or are in swift process of disappearing. Quite a large proportion of this evidence particularly catches the eye of the botanist and ecologist, as for instance the nature and extent of peat digging, the former cultivation of the opium poppy, cole, hemp and woad, and the later part of the book concerns such matters, seeking not only to record the observations whilst they are still to be recollected, but also to set them down against a more general experience of one's own outside the Fenland. Lastly, in face of the extreme rapidity with which the Fens are now altering, I have thought it useful to discuss some of the problems of management of those areas wisely set aside as nature reserves: long service on conservation committees has taught me how complex and involved the preservation of such areas can be; I believe, however, that it can only be achieved by evaluation of the controlling ecological conditions, and a management that conforms to, rather than opposes them.

There are many things this book is *not*. It deals scantily with the northern Fenland and not at all with its coast. It does not describe the detailed history of the drainage of the Fens (although it is necessarily concerned with the consequences), it gives no account of modern agricultural usage of the Fenland, important as this certainly is, nor does it describe the civic quality of the towns and villages of the fen margins and islands. Good authoritative accounts exist on these topics. This remains a story primarily of my own affectionate involvement in the continuing adventure of seeking to understand the course and the processes that have shaped Fenland history and prehistory, sometimes possibly importing hard scientific fact into explanations of fen phenomena that have varied from over-imaginative to fantastical.

2

Ecological background

The progress of scientific knowledge is by no means steady and continuous, except possibly when regarded upon average over a very big field: within individual subjects and areas of enquiry there are long periods of quiescence ended only by the arrival of newer and sharper scientific tools. These tools embrace not merely new techniques and apparatus, but of even more significance, new theories and concepts that may totally alter established lines of thought. This has been conspicuously so with the subject of ecology, which is perhaps more a way of regarding all environmental circumstances and events than a separately defined scientific subject. In these days when powerful lobbies of environmentalists are forces to be reckoned with by businesses and governments and the need for conservation of natural resources is widely accepted, it is extremely hard to believe that even ten years ago 'ecology' was a word hardly understood by the British public, whilst less than fifty years since, biological scientists themselves poured scorn on those few of their colleagues misguided enough to interest themselves in ecology.

It was my own good fortune firstly to have become aware of ecology at school and secondly to have encountered as undergraduate in Cambridge, the great pioneer plant ecologist, A. G. (later Sir Arthur) Tansley. It was in his company, therefore, along with a handful of members of the newly-created 'Botany School Ecology Club', that we made the initial field trip, that took us in 1921 via Upware to Wicken Fen, so beginning a long history of involvement with the subject of ecology as a whole and in particular with the ecology of fen and bog vegetation. This visit was soon followed by field-class excursions from the botanical department by bicycle to the 'Slap-up' at Waterbeach and thence to the river at Baitsbite, from which point we followed the fen road under the river-bank stopping repeatedly to identify unfamiliar plants in the ditches, which at that time were often carpeted with a crimson floating mat of the tiny water-fern, *Azolla filiculoides*, or with the small shining green leaves of the frog-bit. Reaching Upware the ferryman had to be brought over from the 'Lord Nelson', much better remembered as the 'Five Miles from Anywhere'.

Plate 3. The Main Lode, Wicken in summer, 1957. In the deeper water the white water-lily (*Nymphaea alba*), more marginally (foreground) yellow water-lily (*Nuphar luteum*). Much of the fringing swamp is bur-reed (*Sparganium simplex*).

Thence we bicycled by way of Spinney Bank, and Breed Fen Drove and Lode Lane to the entrance to Wicken Sedge Fen and St Edmund's Fen at the head of Wicken Lode. Under the guidance of such able plant taxonomists as Humphrey Gilbert-Carter, Director of the University Botanic Garden, we learned to identify the remarkable wealth of fen plants still growing there. At that time the only substantial ecology carried out at Wicken had been that by R. H. Yapp before the 1914–18 war: he had now moved away and it fell to the post-war generation of ecologists to recommence work on the Fen communities of plants and animals.

What made the study of fen vegetation of such potential significance for the understanding of Fenland history was the fact that plant ecologists had recently established and were applying a very important concept of the behaviour of plant communities. This was the idea of vegetational succession, which recognised that all plant associations, however stable-seeming, were in process of progressive and orderly change. Some of these changes were caused by outside agencies, such as shifts of climate, deposition of fresh soil, different grazing or altered drainage. Apart, however, from these the theory had it that, even in the absence of outside changes, the communities would themselves so alter their own habitats that progressive changes *had* to follow. In terms of wet habitats this meant that one recognised that the colonisation of open water by floating plants and dependent animals leads, however slowly, to the accumulation of organic mud from plant and animal remains which are undecayed in the unaerated bottom layers. This, the 'reaction' of the community itself, causes shallowing of the water, allowing in time the rooted but submerged flowering plants to establish themselves and slightly to hasten the reaction. In time, therefore, we find the reed-swamp community of true bulrush (*Scirpus lacustris*), its leaves below water and flowering stem high above, with the reed-mace or cat-tail (*Typha latifolia* and *T. angustifolia*) and giant reed, *Phragmites communis*. By now it is not only fallen fragments sinking to the lake bottom that are involved but the strong, deeply penetrating stems and fibrous roots of the reed-swamps that are thickening and firming the growing mat of vegetable material. From this time on it is mainly below the surface that the accretion occurs and its pace increases. The material previously formed in open water is best regarded as organic mud, but that which now accumulates is peat, with its fibrous structure and more or less evident derivation from plants growing very close below water-level or just above it. The continuing peat-growth soon allows colonisation by a host of rushes, moisture-requiring herbs and sedges, of which the most robust is the giant sword-sedge (*Cladium mariscus*). Against their competition the last reed-swamp species disappear: the true aquatics have already vanished. As the peat-forming reaction persists and the peaty ground surface emerges, seasonally at first, from the water, the first moisture-tolerant bushes, especially the sallows, establish themselves, to be followed soon by the most moisture-tolerant trees, the alder and hairy birch, which by their greater height can exclude light from lower vegetation, so eliminating the

Plate *4*. Deep-water reed-swamp of *Scirpus lacustris*, the true bulrush: its tall cylindrical flowering haulms project above water, but below water it has narrow ribbon-like green leaves. The white water-lily, its common associate, also is bottom rooted and has submerged leaves.

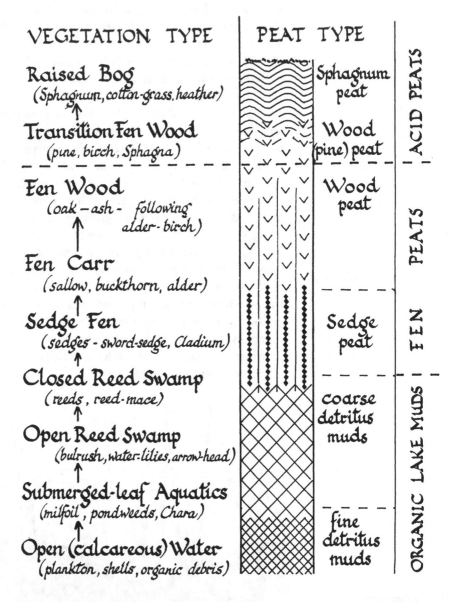

VEGETATION TYPE | PEAT TYPE | ACID PEATS

Raised Bog
(*Sphagnum, cotton-grass, heather*)

Sphagnum peat

Transition Fen Wood
(*pine, birch, Sphagna*)

Wood (pine) peat

Fen Wood
(*oak – ash – following alder – birch*)

Wood peat

Fen Carr
(*sallow, buckthorn, alder*)

Sedge Fen
(*sedges – sword-sedge, Cladium*)

Sedge peat

Closed Reed Swamp
(*reeds, reed-mace*)

coarse detritus muds

Open Reed Swamp
(*bulrush, water-lilies, arrow-head*)

Submerged-leaf Aquatics
(*milfoil, pondweeds, Chara*)

fine detritus muds

Open (calcareous) Water
(*plankton, shells, organic debris*)

PEATS | FEN | ORGANIC LAKE MUDS

Fig. 3. Diagram showing (left) the natural succession of vegetation from open water without outside influence, by the gradual accumulation of the plant remains that it generates: the stratigraphic symbols are those generally used for these organic materials. Note the major change that occurs in suitable climates, when the peats become acidic as they grow above water level: the whole sequence is represented in the Fenland.

sedge–dominated vegetation and replacing it by a wet fen woodland. Even now, the woody roots further consolidate the ground and slowly raise its level so that taller trees like the ash and oak invade and take over dominance. We supposed at that time, that the vegetational succession had reached, with the fen oak wood, a 'climax' stage of stability beyond which, in the conditions of the British climate, it did not naturally progress. We were to find, as Fenland studies of existing and former vegetation continued, that this was an unwarranted assumption and that further stages would lead to displacement of fen woods by yet another and totally different vegetation type. For the moment it suffices, however, to recognise progression as far as fen oakwoods, the position we held in the 1920s.

The self-induced and orderly sequence of events we have described is spoken of as a *primary* succession, to mark it off from those changes induced by outside events. It is an inevitable progression in steady conditions, but it will be apparent how in this particular kind of succession – the *hydrarch* or wet succession – the effect of conditions tending to maintain wetness, such as increased rainfall or worsening drainage, would be to slow it down, whilst conditions making for ground dryness would accelerate it so that the fen wood stage was more quickly arrived at. One might even envisage conditions of such wetness that the succession was set back to an earlier stage. It is not hard to see, in the light of later investigations, how helpful it might be to trace, in the peat of the Fenland, vegetational events that could be interpreted in terms of this concept. However, at the time of Yapp's pioneer research at Wicken he was more concerned with the separation of the plant communities spatially than in a time sequence. Such spatial separation is clearly expressed at Wicken Fen in the zones surrounding such open stretches of water as the lodes and the brick-pits (Plates *3* & *48*). The two sequences are of course controlled by the same basic factor, height of ground-level in relation to water-level, and the zonation from open lode to the fen sedge communities described by Yapp corresponds in general terms with the primary succession we have sketched.

Another concept already accepted by ecologists in the 1920s concerned the source and nature of the water supply responsible for the presence of the waterlogging involved in all the situations and communities in which peat growth occurs. It was already realised that these entities, that today are all called 'mires', include two major types, the fen and the raised bog.

In regions where high rainfall and moisture-laden atmosphere combine, such as most of Ireland and the west of Great Britain, sufficient wetness may be present to permit waterlogging and peat formation even outside the drainage basins. Where these conditions are most extreme, as for instance in western Ireland, Dartmoor and the Pennines, peat may even accumulate over the sloping hill sides as well as on the flat tops and terraces, covering the landscape with a thick mantle of 'blanket-bog', but in somewhat drier climates like those of the central Irish plain, similar mires, essentially dependent directly upon the precipitation, can still develop but are mainly limited to flat valley bottoms, often occupying lake basins after they have been already overgrown (Plates *8* & *9*). Such mires are termed 'raised bogs' or 'domed bogs' for they typically grow into strikingly convex shapes like huge flat sponges, often 2 or 3 miles (3–5 km) in breadth and 30 ft (10 m) or

Plate 5. View from the old windmill on Spinney Bank, Wicken, shewing the boundary ditch and behind it a dense community of mixed sedge, dominated by *Cladium mariscus*, not recently cut and invaded by sporadic bushes. In the background an area longer left uncut and more heavily bush colonised by alder-buckthorn and sallow.

Plate 6. Alder carr at Heron's Carr, Norfolk, with the residue of sedges of the reedswamp and marsh-fern.

more in thickness (Fig. 4). They owe many of their attributes to the fact that they are built of bog-mosses of the genus *Sphagnum* which have remarkable powers of water-retention, thanks to the large water-storage cells which form so much of the bulk of their leaves and stems. Because they depend entirely on the direct precipitation, both blanket bog and raised bog come into the category of 'ombrotrophic' mire; they are also described as 'oligotrophic', since there is so little mineral content in the rain and snow, and the preponderant water excess tends always to drain soluble bases down and away. Small wonder that the growing peat is strongly acidic and lacking in plant nutrients, especially so where the bogs form, as is often the case in the highland zone, upon originally acidic igneous and metamorphic rocks that are hard and slow to weather. The constant wetness, the acidity and low concentration of bases, restricts the plant cover to an extremely specialised and limited range of plants: besides the main peat-forming *Sphagna* and other very numerous associated mosses and liverworts, the higher plants include the shrubby Ericaceae (ling and cross-leaved heath, andromeda, cranberry and whortleberry), the sundews, the cotton-grasses, beaked sedges, some tree sedges and bog-asphodel. They constitute upon the living surface of the raised bog a pattern of pools and hummocks frighteningly quaky to walk upon, and only here and there beside a rivulet or on the curved bog margin does one find a tree, usually pine or perhaps a bush of sweet gale. The more extreme blanket form of ombrotrophic or 'rain-nour-ished' bog supports just the same range of plants but in rather different proportions.

The contrast between these acidic bog-types and the mire category known as 'fen' is extreme. Where peat formation is dependent on the drainage of a large collecting area, the mire type can be spoken of as 'topogenous' or 'landscape dependent' or more generally as 'fen', and it can of course occur where rainfall:evaporation ratios are too low for the growth of ombrotrophic mires. Very naturally the drainage water brings into the collecting basin dissolved mineral salts, especially so in regions of the soft sedimentary rocks where chalk and other limestones figure largely in the catchment area and the water is rich in dissolved bicarbonate and is alkaline in reaction as in the Fenland area of eastern England. Such conditions are

Fig. 4. Measured section through the raised bog at Bettisfield, Shropshire, showing in its peat stratigraphy the stages by which it evolved from open water that was invaded in turn by sedge-fen, fen woods and finally acidic *Sphagnum* bog (symbols as in Fig. 3). Note the gentle dome shape of the bog and the division of the acidic peat into highly-humified peat below and fresher peat above, with pine wood marking dryness at the transition. The dots shew the site of discovery of Bronze Age artefacts. (E. M. Hardy, 1939.)

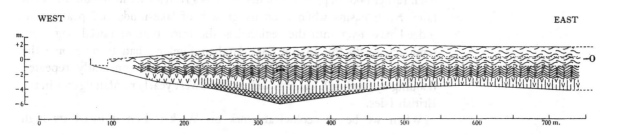

Plate 7. Aerial view of the River Bure, Norfolk: Wroxham Broad left corner communicating by a 'gatway' with the river that winds right round Hoveton Great Broad in which are displayed strikingly concentric vegetation zones. In the open water of the Broad are colonies of floating-leaf aquatics, and the open water is fringed with reed-swamp that in some places is being cut as mowing-marsh. The reed-swamp is succeeded by a bush invasion (mostly by sallows), that leads to tall fen carr dominated by alder, that is tallest away from the open water. This zonation reflects the process of vegetational invasion of the mere after its creation by medieval peat-cutting. (Photograph by Department of Aerial Photography, University of Cambridge.)

intolerable to the plants typical of raised bog: instead the fen vegetation is dominated by other monocotyledonous plants, particularly species of rush, sedge and grass, many of very large dimensions, such as for example the reed-mace, giant reed and sword sedge, robust herbaceous plants like angelica, hemp-agrimony, purple- and yellow-loosestrife, and by such woody plants as the creeping willow, sallow, guelder-rose, the buckthorn, alder and birch. The contrast with the ombrotrophic mire is made the greater since the fen cannot grow upwards beyond the height of the ground-water, save for a few inches, without losing its essential qualities. Of course even in regions of high precipitation, basins that act as drainage catchments begin their evolution by containing fen, though this is often itself rather poor in plant nutrients and has a restricted intermediate kind of flora. Such basins when filled by growth of lake-muds and peat would indeed pass over into the regional acidic mire type of raised bog. Any tendency to do this in the East Anglian Fenland had to overcome the extreme alkalinity of the drainage water and the constantly repeated flooding by base-rich water as well as the lowest yearly rainfall figures in the British Isles.

When we began ecological work on Wicken Fen it owed its high reputation among naturalists to the fact that it still harboured an astonishing wealth of species of plants and animals characteristic of the undrained Fenland, that were preserved here, it is generally held, by the use of the Fen as a catchwater for upland flood-water. The water was held within the clay-strengthened outer bank of its main drainage channel, and maintained in height, by the setting of sluices across the main exit downstream to the river Cam. Collectors, both professional and amateur, botanists and zoologists, found it a treasure house of rare or uncommon species, and indeed they were the benefactors largely responsible for acquiring it as a nature reserve. What we now sought to do was to examine the existing plant communities of the Fen, and arrange them as far as might be in the order of the primary succession. We could recognise the open-water and reed-swamp communities, but only in fragmentary form since the margins of the lodes and brick-pits were artificial and steep; it seemed evident that the next stage was represented by the community of 'pure sedge' dominated by the giant sword-sedge whose growth choked the old boat channels in the Fen and tolerated little competition. It was apparent that this gave place, however, to a bush-dominated 'carr' or scrub of sallow and the alder-buckthorn which progressively thickened until even the giant sedge was killed out, and which itself suffered change in composition as it aged, with successive dying out of the sallow, alder-buckthorn and guelder rose, and their replacement by the purging buckthorn and hawthorn. By this time the

Plate *8*. Aerial view of the great western raised bog at Tregaron, Cardiganshire. It is to a large extent still uncut, although marginal cuttings are visible in the foreground. It is bounded by the River Teifi to one side and by low hills of glacial drift on the other. This bog is 1.5 miles (2400 m) long by 0.75 miles (1200 m) broad and it rises steeply next to the Teifi in a margin 12 ft high (3.7 m): the bog centre is practically flat. (Photograph by Department of Aerial Photography, University of Cambridge.)

Plate 9. The actively growing surface of the raised bog at Tregaron, shewing the hummocks made of *Sphagna*, invaded when old enough by ling, cross-leaved heath, cotton-grass and lichens. In the foreground shallow pool invaded by aquatic species of *Sphagnum* moss, seen breaking the surface here and there.

old carr had a specialised shade-tolerant ground vegetation far different from that of the open fen. There were no established areas of fen wood and it was very remarkable that although alder when planted grew and fruited well, it played no part in the Fen communities; the fen birch was not important, but here and there isolated specimens of oak and ash were flourishing in the carr, occasional saplings occurred and even a young thicket of ash, so that there was the hint that succession might be at the stage of transition to fen oakwoods.

When we had undertaken examination of the behaviour of the Fen water-level through a few seasons and carried out appropriate levelling, it was apparent that the identified communities of the primary succession did indeed grow at ground levels successively higher in relation to the Fen water-table. It seemed that the assumption was confirmed that thus far in its progression the 'reaction' driving the succession along was indeed that of peat accumulation and increasing height: it was clear also that as it had happened in the past, so it was continuing at the present day.

What further emerged from these studies was the fact that much of the Fen was occupied by communities owing their origin to the cyclic cutting of the sedge as a crop for thatch, and to bush clearing in attempts to stave off the bush colonisation of ground that had already become dry enough to support carr. These discoveries, of great interest and potential value for fen management, in no way opposed the concept of the underlying mechanism of the primary fen succession.

The considerable interest of revealing the nature of the primary succession, the 'self-propelled' or 'autogenic' series of vegetational changes, led us to widen the basis of investigation by turning, in the years 1931–33, to the Norfolk Broads, where large areas of shallow open water were surrounded by broad concentric zones of reed-swamp, sedge-fen and fen woods. It was already known by the description of Marietta Pallis (who had little use for the successional concept herself) whilst John Turner (later to become Professor of Botany in Melbourne) and I were also intrigued by reports of acid-loving plants growing on some of the peat lands of the Broads. We concentrated particularly on Calthorpe Broad, small and undisturbed by boating, and now in the ownership of the naturalist, Robert Gurney. Many facts of great interest emerged. Firstly, the zonation of plant communities shewed the expected sequence; the reed-swamp fringing the open water was surrounded by an extensive floating mattress dominated by fen sedges, rushes and grasses (cut when stable enough as mowing marsh). This was evidently invaded by sallow and behind this zone was one of dominant alder and fen birch, an alder–birch carr. Behind this again was a considerable belt of fen wood, containing substantial ash and oak trees of at least three succeeding generations, now all 40 ft (12 m) or more in height: here the alder and birch had been largely suppressed and the undergrowth shewed many species in the ground vegetation characteristic more of woodland than of fen (Plate *10*).

Here we seemed to have evidence that the fen succession, left alone, would indeed progress to woodland, especially since probings shewed us

Plate *10*. Fen oakwood growing upon deep peat at Calthorpe Broad, Norfolk. The mature oaks are 45 to 50 ft (14 to 15 m) tall and represent more than one generation: water-level is quite close to the peat-surface and fen plants such as *Phragmites* and *Angelica* persist in the undergrowth. In the background are bushes of the fen carr.

that this oak woodland was growing over a great depth of peat or mud deposits, so that we could not reach lake bottom. More important still, evidence from the early maps of the broad proved that the zones were not stationary, but that all the zones were marching in towards the open water: thus we confirmed the time dimension of the succession.

We also found that in many places in the alder–birch carr there was a discontinuous growth of a carpet of *Sphagnum* moss of two or three different species, together with the hair moss, *Polytrichum* sp., and we noted the very successful growth of planted Rhododendron bushes, both clear indications of acidic soil conditions. When we had made systematic measurement of level and of soil acidity along many lines radiating outwards from the lake it became clear that as the peat level was raised progressively at increasing distance from the broad, so we passed from fully alkaline open water and fen, to neutral and then to decidedly acidic soil reaction at the sites where the *Sphagna* were establishing themselves. This we took as an indication of a possible tendency for the fen succession to progress from the fen wood stage into growth of acidic raised bog. Some hint that this might be so lay in the successful growth of scots pine planted in the Fen oakwoods, for this tree is typical of the natural transition phase to acid bog. By good fortune it happened that by this time we had begun work upon the peat deposits of the southern Fenlands and that a site between Ely and Littleport was yielding clear evidence pointing in the same direction: how these stratigraphic enquiries originated and developed are, however, separate matters to be pursued in later chapters. It will already be evident that our ecological studies had by now provided a promising route of enquiry into the former natural conditions of the peat-fens.

3

Pollen analysis

In 1931, following a suggestion from A. G. Tansley, my wife began research in pollen analysis, a technique recently brought to notice by papers written in English by the Swede G. E. Erdtman, who reported some preliminary studies of peat deposits in the British Isles, and gave an account of the methods developed by Lennart von Post. Broadly speaking the method at this time depended on the principle that since the ice sheets retreated, the natural vegetation of western Europe has been forest, the trees of which are mostly wind-pollinated and have produced pollen in such vast amounts that in any open area the yearly rain of pollen on land, water and foliage has been considerable, amounting to many thousands on each square centimetre. When the pollen grains have fallen into a waterlogged environment, such as a growing peat-bed or lake bottom, they have escaped decay and have become incorporated year by year in the growing deposit. A large proportion of the grains have morphological characters such as the number, size and construction of the pores, the thickening and marking of the surface, overall size and presence or absence of air-sacs, and although the protoplasmic contents swiftly disappear the outer wall, the exine, is exceedingly resistant to decay, and to many chemical reagents. Thus the grains can readily be identified microscopically in a sub-fossil state after suitable and perhaps severe treatment to free them from the mineral or peaty matrix that contains them (Fig. 5).

From a very small sample, therefore, of peat or lake mud, no bigger perhaps than a wheat grain, one could identify some hundreds of tree and pollen grains, and from their relative proportions have an index to the relative frequency of the various trees in the region's woodlands at the time of formation of the sample. The possibilities of the method extended far beyond this, however. It was possible by comparing the tree pollen frequencies in a consecutive series of samples through a deposit, to discover the alterations in forest composition throughout the period of accumulation. So far of course, as the forest composition had been determined by climatic causes, so the changes in forest history also reflected the progressive alteration of climate. What gave the method remarkable strength was the

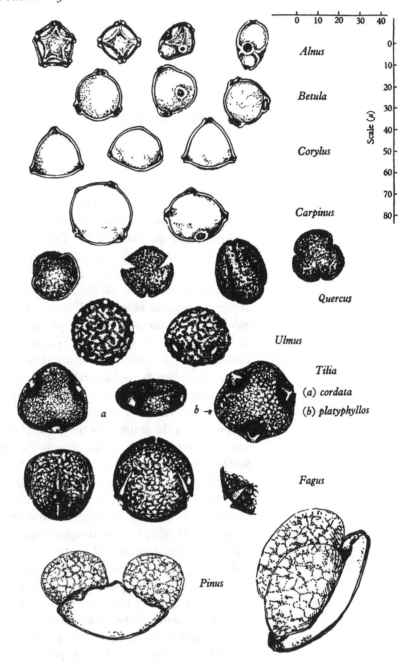

Fig. 5. Drawings of the chief types of tree pollen found in British Flandrian deposits, drawn to a common scale. A very wide range of other pollen types of shrubs and herbs is now regularly identified and counted. In descending order the genera are alder, birch, hazel, hornbeam, oak, elm, lime (*a*) winter (*b*) summer, beech and pine.

Alnus

Betula

Corylus

Carpinus

Quercus

Ulmus

Tilia
(*a*) *cordata*
(*b*) *platyphyllos*

Fagus

Pinus

Scale (*µ*)

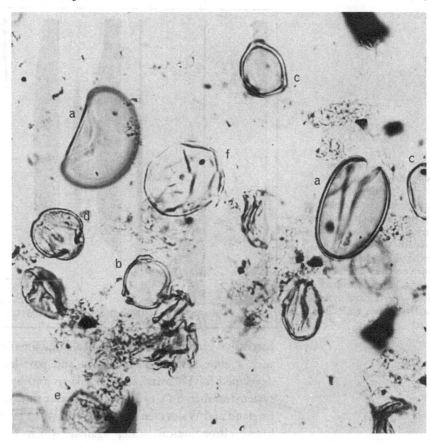

Plate *11*. Pollen grains and spores as seen under the microscope in a preparation of a typical peat or aquatic mud. The largest, bean-shaped objects (*a*) are fern spores: most of the others are pollen grains of trees, (*b*) birch; (*c*) hazel; (*d*) oak; (*e*) elm; the large crumpled grain (*f*) is probably grass pollen. As in most fen peats, there are also numerous corroded and broken grains. Magnification about ×450. (Photograph by R. G. West.)

demonstration by the pioneer Swedes that, in sites distributed through quite a large region, the pattern of forest history had been closely similar and indeed over western Europe as a whole a very broad but generally consistent pattern of forest evolution could be proved. Although not entirely so, this consistency was based upon the fact that the forests were responding to the same widespread climatic changes, advancing in their broad latitudinal belts northwards as climate became progressively warmer with the melting of the European ice sheets, attaining indeed at one period, the so-called 'climatic optimum', latitudes further north than those they would naturally occupy today. It seemed from the measurement and comparison of the annual laminations in the deposits of lakes abutting on the melting glaciers that the great and final retreat of ice had begun in central south Sweden some ten thousand years ago, and now we had in prospect a time-scale to cover changes through what was then spoken of as the 'post-glacial' period. Von Post applied to his south Swedish pollen diagrams a numbered system of zones and by appropriate local pollen-analytic studies demonstrated how

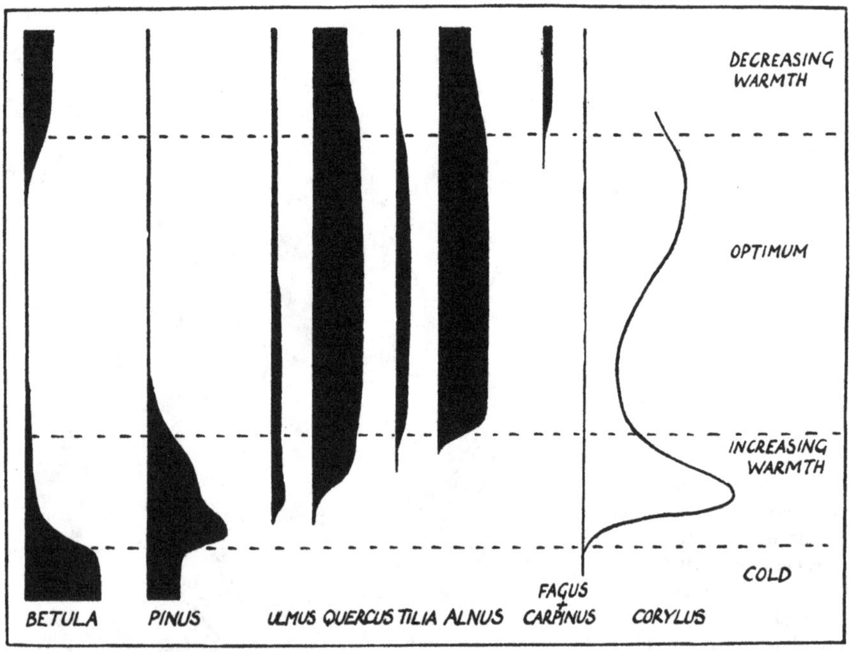

Fig. 6. Generalised form of pollen diagram from East Anglia. It represents the changes in forest composition through the whole Flandrian period to the present day. From the left the tree genera are birch, pine, elm, oak, lime, alder, beech and hornbeam, hazel. The frequencies are given as percentages of the total tree pollen, excluding hazel.

successive geological, archaeological and climatic events might be tied into such a zone system. Comparable and parallel zone systems were soon developed for Denmark and Germany, and by 1940 the outline of a zone system numbered IV to VIII for the same time span had been published for England and Wales (see Figs 13 & 18). What in fact we had arrived at was a broad chronological scale, against which to measure every kind of happening through the last ten thousand years. In the period before absolute physical means of dating were available, the importance of such a background means of correlation and reference was immense, and to the extent that we could match Scandinavian post-glacial events with those in this country, we now had a tentative scale in years derived basically from the lake lamination studies of Southern Sweden and Finland. In practice it was common to refer events to one or other of the sequences of '*Blytt* and *Sernander*' climatic periods, more as a guide to age than as indicative of the past climatic conditions. In 1931 the significance of this scale was roughly that shewn in the table representing southern Scandinavia.

Having learned for ourselves how to extract and identify the chief types of fossil pollen, it was not long before we were able to apply pollen analysis to geological and archaeological situations. We began with a series of peat beds between thicknesses of coastal clay disclosed in trial excavations in the inshore part of Swansea Bay and referred to us by Professor O. T. Jones, Sedgwick Professor of Geology. We were delighted to discover that these beds had distinctive tree-pollen composition and might by this means be referred to appropriate climatic periods. The lowest fresh-water peats sandwiched in the clays now 50 ft (15m) below sea-level were of middle Boreal Age, those about 20 ft (6 m) below sea-level were late Boreal, whilst the samples from about −5 ft (1.5 m) and above belonged to the Atlantic period. We had therefore been able to shew that a rapid rise in sea-level of 50 ft (15 m) or more had been in progress in the Boreal and perhaps early Atlantic time and had brought the ocean to its present height. The relevance of this to the East Anglian Fenland is not at first apparent, but a big rise in ocean level in post-glacial time was well known in Scandinavia, and indeed was to be expected if we recall that the melting of the great ice sheets of the world must have restored a great deal of water to the oceans. A figure suggested by calculations from the former extent and thickness of the ice was some 300 ft (90 m) for the rise in sea-level throughout the world. This was the so-called great *eustatic* rise, and its dimension was such that before it occurred the bed of the North Sea must have been above sea-level, a great lowland plain across which flowed the lower courses of the Rhine, Thames and Trent and which afforded dry-land access to the British Isles from the Continent for plants, animals and prehistoric man: it was clearly going to be of great importance to learn the time of this 'dry-land phase' of the North Sea and of the time when the rising ocean level cut us off from mainland Europe. We very soon found ourselves more directly involved with this problem.

It had been known for many years that trawlers fishing on the Dogger Bank had their nets fouled by a brown fibrous layer known as the 'moorlog': large detached lumps proved to be a fresh-water peat often yielding recognisable plant remains, including those of the arctic dwarf birch. Early pollen-analysts shewed that samples dredged up from depths of 120 to 180 ft (36 to 55 m) below sea-level had apparently formed in the early Boreal

Early chronological scale: southern Scandinavia, 1931

Climatic period	Character	Forests
Sub-atlantic	Cool and wet	Spreading beech, spruce
500 B.C.		
Sub-boreal	Dry, warm and continental	Mixed oak forests
3000 B.C.		
Atlantic	Warm, moist and oceanic	Mixed oak forests
5600 B.C.		
Boreal	Dry, warm and continental	Hazel scrub, pine
7500 B.C.		
Pre-boreal	Cool summers, severe winters	Birch
8000 B.C.		

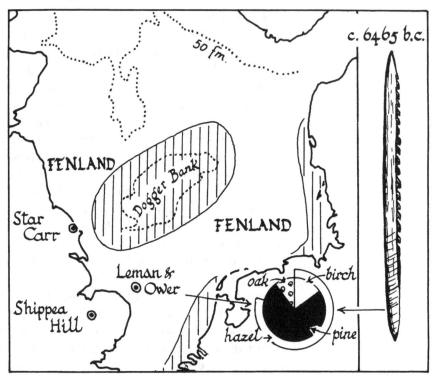

period. This evidence was dramatically supplemented when, in the summer of 1929, the skipper of the trawler 'Colinda', fishing between the Leman and Ower banks off the coast of Norfolk dredged up, from a depth of about 120 ft, a piece of moorlog which, on being broken open, proved to contain a 'harpoon', a worked bone point some inches long, serrated on one side with well fabricated teeth, of a type known from Estonia and associated by archaeologists with a Danish Mesolithic (i.e., pre-Neolithic) culture. The skipper had unfortunately jettisoned the lump of moorlog but on two later occasions in 1932 recovered from the same area samples of moorlog for pollen analysis. Our analyses (and those independently done by Erdtman) shewed the moorlog to have formed in the Boreal period, the age also attributed to the Estonian and Danish discoveries (Fig. 7).

The Fenland basin is in fact no more than the shallow margin of the extensive North Sea plain, and it remained to be seen how its history related to the submergence that now occurred. We had no need to pause, for early in 1932 we had been invited to see the unusually deep and extensive excavations that accompanied the building of a new sluice at Wiggenhall St German's, 4 miles (6.4 km) south south west of King's Lynn, across the outfall of the Middle Level drain into the tidal Ouse. It extended to 23 ft (7.5m) below Ordnance Datum and disclosed four peat beds at different

FEN BEDS AT ST. GERMAN'S NR KING'S LYNN.

	MADE GROUND	DEPOSITS	FORAMINIFERA DERIVED	INDIGENOUS	OTHER INDICES	CONDITIONS
		J Blue Clay	few	more than in G		Estuarine to Marine
		H Peat				Fresh-water
		G Brown silty Clay	fewer than in F	fewer than in F		fresher than I
		F Blue Clay	from the Chalk frequent	abundant, species & individuals	Scrobicularia piperata	Estuarine to Marine
		E Peat			Glass beads La Tène to A.S.	Fen-woods ↑ Salt-marsh
		D Blue Clay	few	few species and individuals	Cardium edule & tibia of a deer.	Brackish Lagoons
		C Peat (locally pebbles)				Fen-woods
		B Blue Clay	from the Kimmeridge, frequent	few species & individuals		Brackish to freshwater
		A Peat				surface soil
		Kimmeridge Clay				

Depth scale (left): +6, 5, 4, 3, 2, 1, ORDNANCE DATUM, 1, 2, 3, 4, 5, 6, 7, 8, 9, 10, 11, 12, 13, 14, 15, 16, 17, 18, 19, 20, 21, 22, 23, 24 — M. FEET

Fig. 8. Sequence of Flandrian deposits revealed in the deep sluice excavation at St German's, showing the results of foraminiferal analyses by Dr Macfadyen and the implications as to conditions that they indicate. The lowest (fresh-water) peat bed is now over 23 ft (7 m) below sea-level, the next highest has rooted oaks *in situ*, and the one above that also carried fen oakwood (see Fig. 10).

depths separated by clays and silts laid down in brackish or salt-water conditions (Fig. 8). Even the lowest thin peat sitting on the surface of the Kimmeridge Clay shewed by its tree-pollen content that it was post-Boreal in age, as naturally were also the higher peat beds, so that it was evident that most of the great eustatic rise in ocean level was over by the end of Boreal time. Whilst the pollen analyses could not date the peat beds more closely they taught us an extremely valuable lesson, thereafter held in mind in all our Fenland investigations. The third peat bed from the surface, though only 6 in (14 cm) thick, contained abundant remains of prostrate oak trees, now compressed to oval section 4–8 in (10–20 cm) wide, and stumps *in situ* shewing by their horizontal root-systems confined to the peat layer that they had grown as a horizontal plate in the aerated layer above the fen water-table, just as we had seen with the living oaks at Calthorpe Broad. Moreover, the pollen series through the bed shewed that in the lower layers alder pollen strongly preponderated, whereas higher levels were totally dominated by oak pollen. Such pollen diagrams are not easily to be related to regional forest history for they clearly reflect no more than the local fen succession and this example represented the replacement of the alder carr by the fen oakwood. Evidently further flooding by sea-water put an end to this fen succession. The next higher peat-bed, 2 ft (0.6 m) in thickness, analysed with these conclusions in mind, beautifully illustrated a fuller vegetational succession. At the base there were maxima of pollen highly typical of coastal salt marsh, these gave place to types indicative of fresh-water fen with strong early dominance of alder and willow (carr) and then still higher the oak increased so as to dominate all else (Fig. 9). At this stage fern spores were very abundant indeed and there can be no doubt that this represents the fen oak woodland stage. At the very top pollen of salt marsh species indicates the approaching return of brackish water conditions and indeed the bed is now grown through by stems of the salt-tolerant reed. Such an actual vegetational succession may still be seen here and there behind the coastal salt marshes of northern Norfolk, and it was apparent that we may often expect the local fen successions to express themselves strongly in analyses of the Fenland peats and that we should indeed seek to take advantage of this possibility.

Very soon we were presented with a remarkable opportunity to develop this kind of enquiry very productively. Between the north-east extension of the Isle of Ely, in the small elevation of Brick Hill, and the upland of

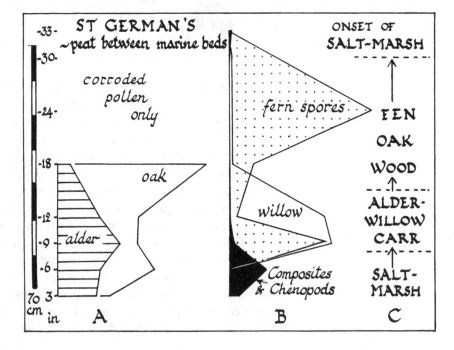

Fig. 9. Pollen analysis through a 30 in (76 cm) peat bed, at St German's, near King's Lynn, between marine clays (see Fig. 9) and now 3 to 5 ft (1–1.6 m) below mean sea-level. *A*, relative frequency of alder and oak as percentages of total tree pollen; *B*, absolute pollen frequencies of salt-marsh herbs, shrubs and fern spores; *C*, interpretation of the pollen data in terms of the local succession of vegetation. A similar peat bed lower in the sequence actually had oak stumps *in situ*.

WOOD FEN , ELY   1874

Fig. 10. The 'forest horizons' in Wood Fen, near Ely described in stratigraphic position a century ago by Miller and Skertchly. The original high forest of tall oaks was rooted in the basal clay: later trees are smaller, rooted only in the peat and have horizontal root systems. These trees are mainly pine, yew and oak, one generation of tree often seen grown astride the fallen trunks of an earlier one. An uppermost horizon of sallow and alder has been omitted.

Littleport there is a triangular area of fen, known as Wood Fen or North Fen, made famous by the descriptions of the observations made by W. Marshall of Ely and Skertchly and published in 1877 and 1878, at the site from which no less than five horizons of buried forests had been recognised. We now learned that farming operations were bringing to light once more the trunks of large trees, and local excavations were arranged by Gordon Fowler who was specially concerned to discover the causes of forest destruction. What Marshall had described was (1) a lowest layer of large oaks with large prostrate stems and stools rooted firmly in the basal clay of the fen floor, (2) a layer immediately above of somewhat smaller, but still large trees, rooted in the fen-peat with horizontal root-systems and sometimes actually growing directly upon trees of the layer beneath, (3 and 4) two layers of large pine trees again with both large prostrate trunks and stools with horizontal root plates and sometimes bestriding older fallen trunks, (5) an upper layer of alder, willow and sallow (see Fig. 10). It proved quite easy to confirm from the timber still present the general sequence as described, except that peat wastage seemed to have destroyed the uppermost tree layer. In the layers where the pines grew there were some well-preserved ripe cones and the characteristic pine 'needles', short shoots bearing two leaves, and along with the yews were their typical pollen-bearing cone-scales: it was absolutely clear that both had grown quite naturally in past peatland communities.

Pollen analyses from a series of samples taken where the peat was deepest proved, as at St German's, strongly to reflect the local conditions. Although alder pollen was abundant throughout, the lower half of the series was dominated by oak pollen (reflecting forest layers 1 and 2), and the upper half by that of pine (layers 3 and 4) although there was no evidence to suggest the

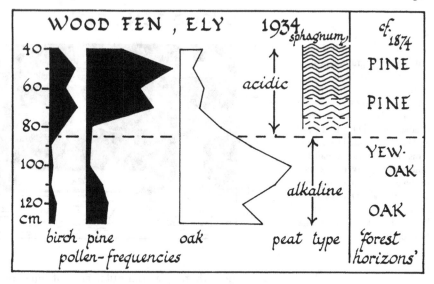

Fig. 11. Diagram to show correspondence at Wood Fen between the described 'forest horizons', the development of acidic *Sphagnum* peat and changes in frequency of the three tree-pollen types essentially indicative of the early high forest and calcareous fen (oak) and of the developing acidic conditions (birch and pine).

need to divide the pinewood phase into two. The uppermost pollen sample shewed such high values of alder as to suggest that it might be the last trace of the uppermost alder–sallow–willow layer of the series. Examination of macerated peat from nearby sites in the Fen fitted closely to this sequence, there were oak leaves in the lowest peat, and we recognised that alder timber, twigs and cones tended to be present throughout, although less frequently in the pine phase, where, however, fruits of tree birches were abundant. The macerated peat samples also yielded important evidence of the herbaceous vegetation. The large stems of the giant reed (*Phragmites*) and roots and fruits of the sedges were present throughout the two oakwood layers, and the indication of alkaline fen was supported by the local presence of *Chara*, a large green alga known as stonewort from the deposition of chalk over its surface to form a soft concretion. It was quite other in the pine tree layers above; here, and especially towards the top we found the frequent stems and leaves of *Sphagnum* moss, a clear indication of ground acidity, and this was accompanied in the pollen counts by high frequencies of the spores of this moss (Fig. 11).

The evidence thus very briefly sketched clearly related to the tendency we had observed in the present-day fen succession at Calthorpe Broad, and suggested very strongly that we were dealing with the natural development from alkaline fen through the fen oakwood stage by increased acidification towards the genesis of raised bog. Very fortunately this vegetational succession is well represented along the flat coasts of the Baltic where in the region of the *Kürische Haf* the fen vegetation growing by the almost fresh water can be seen giving place inland to extensive raised bogs dominated by *Sphagna* and essentially maintained by direct precipitation of rain and snow

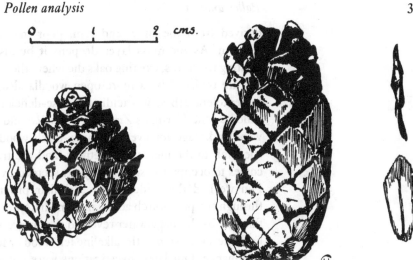

Fig. 12. Cones and cone-scales with the marks of the pair of attached winged seeds, recovered from the pine stub layers at Wood Fen, near Ely.

upon them. The fen successions proceed as already described up to the stage of fen oakwoods but as the peaty surface is raised sufficiently (and very few centimetres will do) to escape flooding by the alkaline ground water, downward leaching of bases by precipitation and the accumulation of acidic plant remains continue. These soil conditions and freedom from desiccation allow *Sphagna* to enter, and as this very water-retentive moss carpets the ground, the surface is once more waterlogged, although now by water

Plate *12*. Stools and split trunks of pine (*Pinus sylvestris*) taken out of Wood Fen, near Ely, and waiting disposal. They represent a local phase of acidification of this Fen, probably in the Sub-boreal period (see text of Chapter 4).

derived from the snow and rain, poor in plant nutrients and acidic in reaction. As the moss layer deepens it builds the water-table above the existing tree roots, existing oaks die where they stand and do not regenerate, but are replaced by a more open woodland of pine and birch, trees that better tolerate these wet acidic and base-deficient surface conditions. This is called by the Germans *Zwischenmoorwald*, and it is indeed an intermediate woodland stage for even the pine–birch wood is destined before long to succumb to the increasing wetness, and the treeless dome of a raised bog comes to occupy the site (Fig. 3). It seemed extremely likely that at Wood Fen we had the evidence for a similar progression from fen oakwoods through the pine–birch acidic intermediate woodland, after which renewed flooding by drainage water reversed the progress towards raised bog, setting back the succession to the alkaline fen wood or fen carr stage of alder–sallow dominance. Our later investigations were not only to confirm the truth of this interpretation but to shew that in some marginal parts of the Fenland basin the natural progression had advanced through to the stage of fully developed raised bog that was active and carried its typical plant communities until destroyed by the last drainage improvements of the middle of the nineteenth century.

The regional tree-pollen disclosed by the pollen analyses made it apparent that, as at St German's, all the peat deposits were of post-Boreal age, so disposing of earlier conjectures that the lowest forest layer might reflect the 'warm dry' Boreal period. The presence of both beech and hornbeam pollen in the uppermost layers gave some indication that they formed late in the zone sequence, perhaps in the Sub-atlantic period. It was quite apparent that local fen vegetation had strongly influenced the Wood Fen pollen diagrams and this was a feature that we had to reckon with throughout all our later investigations of the Fens. In order to make best use of the pollen-analytic method to provide a time scale, the local influences had to be discounted so that we could discern the broad regional forest history of the uplands, but on the other hand the progression or setting back of the natural fen successions indicated by the local pollen could be used to tell us of the past changes in wetness of the Fens that were so much controlled by marine ingression or regression even when the salt water did not reach the peat fens themselves.

4

Bog oaks and buried forests

As the decayed remnants of a Roman villa or the isolated circle of monoliths on a bleak hill top never fail to evoke an awareness that past civilisations have been present on our own familiar countryside, so through the centuries the sight of massive forest trees below the now treeless Fenland has created in their discoverers a strong sense of contact with the landscape of former times when things were not as now. Giant trees have in fact been disclosed in vast numbers by drainage cuts, peat-cutting or wastage, throughout the peat Fens, more especially in a belt a few miles wide extending seawards from the margin of the Fens against the upland.

They are referred to collectively by the old fenmen as 'bog oaks', but among even the biggest trees yews and pines are well represented. A rooted yew I saw excavated in Isleham Fen in 1935 (Plate *13*) had a girth of 14 ft 6 in (4.4 m) and its wood was still exceedingly hard: indeed with the yews, as with many of the oaks, a heavy felling axe striking the timber squarely, will often rebound rather than cut. With the oaks this may well be associated with the deep black colour extending through the timber and possibly due to soluble iron from the mineral sub-soil reacting with tannins just as in the ancient process of making writing ink. Despite its hardness in the field much of the timber is of little use when extracted, since upon drying it develops great 'shakes', i.e., splits that extend throughout and especially along the wide medullary rays that are so typical of oak and give its attractive grain to near-radially cut and polished surfaces. Even for fuel it is by no means a total success for it is very hard to cut up, tedious to dry and contains too much ash to burn well; thus one often sees the great crowns and boles of the bog oaks lying beside the ditches where they decay slowly year by year and often shew evidence that attempts to burn them have failed. However, in harder times they were not neglected. The very experienced fenman Will 'En Edwards of Lotting Fen has described how one or two enterprising contemporaries about the turn of the century avoided the cutting whilst taking advantage of the slow burning of the bog oaks. A straight oak trunk was introduced into cottage or public bar occupying its whole length, one end suspended by tackle to burn in the fire place, the trunk providing seating for family or

Plate *13*. Stumps of a large yew (*Taxus baccata*) exposed at Isleham Fen, Cambridgeshire, being examined by Professor A. C. Seward, Chairman of the Fenland Research Committee, and the author. The tree is part of the buried high forest rooted in the Gault Clay, and is one of many such giant trees of oak and yew taken from this Fen about 1935 when this picture was taken.

customers: it was jacked forward day by day as the combustion slowly continued. He also drew attention to the great quantities of timber taken from the Fen, citing an advertisement of a parcel of 300 tons of 'Black Oak' for sale from a single small farm. The resinous pines make better fuel and the yews, less liable to 'shaking' because the rays are so fine and because resin-canals are lacking, survive drying better and are sometimes used for wood-turning of small articles which when finished have a satiny surface and a warm sherry brown colour.

Of course conjectures of the age of the buried forests and the causes of their destruction have been numerous and wild. As one would expect, Noah's flood was often invoked, and, in more general terms and not altogether wrongly, a general rise of sea-level, as seems very clearly evidenced by the forests in the peat beds exposed by erosion over many parts of the British coast line. A great catastrophic gale has been thought to be indicated by the common general direction of the fallen trees, a majority of which do indeed lie with their crowns to the north east, but as at Wood Fen, the buried trees clearly fell at very different times. I am not impressed by the idea that the trees fell in one direction simply because of the wind-shaping of the crowns by the prevalent wind, for every individual tree grows mechanically balanced so exactly that it may readily be felled by choosing where to cut the bole, and with sapling trees if one weights branches on one side, all the others respond by changes in size or direction until the gravitational equilibrium has been restored. This response is little investigated or appreciated but its consequences can be seen everywhere in modified tree form following lopping and shaping. In fact the mean direction of south-westerly gales over a longish period of time is the most likely cause of the common direction of fall.

It has often been pointed out that stools of the fossil trees still rooted in the solid fen floor often shew curiously conical apices to the stumps, as if they had been felled and often they look to have been charred. In fact it is extremely hard to prove, one way or another, if prehistoric man had a hand in cutting or firing the forest. In all his role must have been a small one, for we now know that the great forest of the fen floor was generally not later than the Neolithic age when man's cutting tools were at best only polished stone axes. Moreover, the effects of charring upon standing stumps are often impossible to distinguish from those of decay in peaty conditions, and during the drying out of the Fens the upper peat layers have often suffered very persistent peat fires that can eat down well below the surface, so that the carbonisation may long post-date the death of the trees. In a mood of enthusiasm and seeking to resolve this problem, Gordon Fowler once publicised among the fen farmers an offer to give a golden sovereign to anyone who could give him the exact locality where buried trees shewed indisputable signs of having been burnt; responses were so quick and numerous that the offer was hardly made before it had to be withdrawn, but they yielded no decisive answer. Happily most of the conjecture has been resolved as more systematic and scientific enquiry has proceeded.

BASAL HIGH FOREST

What has first to be recognised is that the buried forests are of two distinct types of differing character and causation, although buried woods can now and again have a somewhat intermediate character. The first of these is the primeval high forest that before the advent of Neolithic man covered the whole of the British Islands, except the northern highlands of Scotland, mountain tops, coastal marshes, bogs, lake basins and river banks, with a mantle of tall deciduous trees, whose continuity is hard for us to realise at the present day. To a large extent it was dominated by oak but wych elm and

Plate *14*. Trunk of a giant bog oak taken from the peat at Queen Adelaide Bridge, near Ely, in 1960. Its size and un-branching straight bole (70 ft (21 m) in length) are typical of the basal buried high forest, and its radiocarbon date of 2535 ± 120 b.c. confirms this. (Photograph by J. Slater, Ely.)

the two native lindens were also present. Locally on the poorer sands and gravels there were tall stands of the pine with birch, able in these situations, as on some marginal bog sites, to avoid the competition of the broad-leaved trees. There were frequent depressions, such as the North Americans call 'swales', where such trees as alder and willow, tolerant of high levels of water in the soil, formed wet glades, whilst wherever ancient trees collapsed, fell and decayed in the forest, the gap so created was colonised by quick-seeding pioneer trees like ash and birch, forming local thickets that slowly thinned as the taller dominant trees grew to fill the gap. In such temporary clearings and on the forest edges hazel was abundant. The yew had considerable powers of persistence within the full forest cover, and one suspects that it favoured the margins of the alder glades where the wet soils were becoming peaty.

In the severe competition for light in the high forest, lateral tree branches soon die and fall away so that the majority of the trees have straight unbranched boles carrying surprisingly small and compact crowns of foliage. This then was the aspect of the woodland that is represented by the basal layers of buried trees throughout the Fenland. The remarkable straightness and lack of branching in the prostrate trunks have been often commented upon, and their large dimensions have provoked still more conjecture. The bog oaks are not uncommonly as much as 90 ft (27 m) in length without branching, indicating a general forest height far greater than that in existing British oak woods (Plate *14*). Although they were growing in a climate somewhat warmer than the present, it is not necessary to suggest this as the factor primarily responsible. In the one great European natural high forest still remaining, part of the Czar's former hunting preserve in eastern Poland, the forest trees, including the limes and oaks in large amount, generally attain heights of 120 ft (36.5 m) and have small crowns and long unbranched trunks like the fossil trees already described. We no longer have trees of these dimensions in Britain, in large part because our forests have been so long and heavily exploited. It seems probable that elms were selectively destroyed even by Neolithic man, and that the limes with their useful bast fibre and easily worked wood suffered similarly. The oaks themselves, it now appears, have for centuries been very strictly managed in the small areas of residual forest: not only were the tall standard trees regularly cut, but all other oaks were regularly coppiced and cut for small building timbers as the stems reached 30 or 40 ft (9–12 m) in height. It would be surprising indeed if this removal of timber continuously through at least the last nine centuries had not in the end seriously depleted the forest soils of critically important nutrients, such as phosphates. This is a process exactly analogous with the impoverishment of Welsh mountain pastures by 'the carrying away on four legs' of soil phosphates after the introduction of intensive production of lambs instead of mature mutton.

DEATH AND ENTOMBMENT

It is intriguing to realise that although natural high forest no longer grows in the British Isles we have entombed below the Fenland peat a sub-fossil example of its impressive grandeur. There remain for consideration the cause of destruction and burial of the basal forest and the time of this very important event in the evolution of the Fens. Over the southern half of the Fenland, where the bulk of buried trees have been found, careful excavation will reveal, where chance does not already shew it, that the roots of the big basal trees are firmly rooted in the solid clays of the fen floor, mostly Gault, Kimmeridge or Oxford Clay or in Boulder Clay largely derived from these formations: penetration into such heavy soil must indicate that the forest was then quite free from persistent water-logging such as is requisite for peat formation. The death and fossilisation of the forest were caused by the onset of waterlogging throughout the Fenland basin. Although the water was not very deep it brought about extensive formation of fen and marsh, below which the forest soil was totally deoxygenated so that the roots in it could not respire and the forest trees quickly died. Like the tree stools, the heavy fallen trunks of the dead trees were soon submerged in the accumulating peat where again the anaerobic conditions prevented their decay. At the junction of air and water the stools were specially subject to fungal, bacterial and insect attack, leading to quick collapse of the heavy trunks, especially of course in gales, and to the conical shaping of the remaining stump. In the area of heath and woodland centred about the region of Thetford, Bury St Edmunds and Mildenhall, the ice sheets of the last glaciation crossed the outcrop of the Greensand so that the glacial drift is extremely sandy in character, and in a substantial part of the eastern Fenland adjacent to this 'Breckland' the sub-soil is of similar character. In this area that includes Isleham and neighbouring Fens, the basal forest contains large and abundant yew and pine in addition to oak. Whilst there is no doubt that this difference in forest structure is due to the soil difference we can point to no present examples of living forest with which to make comparison. Be that as it may, there is no doubt that the destruction of it was part of the same phenomenon as that of the forests on the heavy clays.

A proper understanding of the causes of the general waterlogging that initiated the widespread waterlogging of the Fenland high forest has involved the use of many scientific techniques and the collection of much field information. The early stages of enquiry were particularly handicapped by the lack of chronological indices and dating was largely dependent on the infrequent discovery of archaeological objects associated with the fen floor. Burwell Fen was a particularly useful source of evidence, since in the nineteenth century big areas of peat were stripped away to give access to the surface of the Cambridge Greensand, from which coprolites, the primary raw material in the new phosphate industry, were extracted. Polished Neolithic axes and a grinder made of sandstone suitable for their manufacture, together with fragments of the foreign rock from which they were made, were found there. Objects of the later Bronze Age cultures were also found, but one assumes it was the lowest forest layer that was specifically associated with the earlier of the two cultures.

There is good reason to believe that throughout these parts of the fens the general waterlogging was caused by the backing-up of fresh water from the rivers during the last stages of a rise in sea-level relative to land-level. If this were so we should expect the ages of trees from the basal forest layer to be broadly the same, but until the advent of radiocarbon dating we had no means of closely testing this hypothesis.

FEN WOODS

Impressive as the trees of the buried high forest are, they by no means represent all the buried timber of the Fens. A second category of woodland community was at least equally characteristic and widespread, the indigenous fen woods, examples of which still may be found in many parts of the country, colonising fens where the natural vegetational succession has given rise to peat growth sufficient for it to attain the mean water-level and afford those first footholds that can be colonised by woody plants such as sallow and alder. More particularly the alder is characteristic of such situations and as it grows so much taller, will indeed soon become the dominant tree of thickets perhaps 40 or 50 ft (13–16 m) high. It has the great advantage that it has a system of roots that grow steeply down into the unaerated peat affording good anchorage even in these unstable situations. The alder stools are often many-stemmed and constitute small islands upon the weak peat surface: in high winds one may sometimes see individual trees swaying upon their separate tussocks. In such conditions the trees tend to sink under their own weight and if for this cause, or through worsening drainage, water-level goes on rising generations of alder trees may be embedded in the growing peat. A great deal of the peat of the Fenland is of this kind. At Wicken Fen and round about, the sedge-peat to a depth of more than 12 ft (3.6 m) is full of the twigs and very recognisable cones of the alder and of fragments of the deep crimson wood with its black shiny bark crossed by long lenticels somewhat resembling that of hazel. The smaller fruits are easily washed out also and the very recognisable pollen grains are present in such immense numbers as to eclipse altogether those of other trees. Despite this undoubted prevalence of alder, substantial remains of its timber are not very frequent, one supposes because the stems decay so quickly on exposure to aeration. The same applies to the sallows and to the fen birch which often is an important component of the fen woods, although the latter again is readily identified by its fruits, and at generic level by its pollen, as with caution, also by its wood and bark.

The dead fallen stems of the fen wood trees and more especially their buried root-systems contribute to make a type of wood-peat that is very recognisable, especially when examined in the laboratory, and increasingly with time the peat becomes compacted and grows higher, so that in the absence of further external flooding, fresh stages of the vegetational succession establish themselves. Now trees more susceptible to ground flooding may enter to compete with alder and birch, especially the oak, but also to some extent ash, yew and, in certain circumstances, pine. These are all trees of greater height and longer life that will totally replace the earlier occupants; as we have seen, they may grow to considerable size though much less than that in the buried high forest (Plate *10*). Their root-systems too are now in the form of a shallow horizontal plate confined to this shape by a water-table just below the ground surface, and where such fossil woods are excavated it will be seen, as at Wood Fen, how often the roots of one generation of trees have found support across the fallen trunks of an earlier one. All the trees so characteristic of these mature fen woods occur as buried timber in the Fens. It is not surprising to find the ash, although it is badly affected by poor soil aeration and develops severe bacterial canker in these conditions. The frequency of yew is more surprising for it has been regarded as typical only of well drained and generally limestone soils. My own attention, wakened by the numerous fossil discoveries of the tree, soon noted that in many derelict plantations and half-wild copses on the Fens, it grew extremely well on the alkaline peat as also in the somewhat more natural fen woodlands at Chippenham Fen just north of Newmarket. I realised too that in the buried tree layers in the peats of the Thames estuary yews had been often reported, and certainly from our Fenland sites we recovered not only timber but the hard seeds (shaped like small hazel nuts) and even the scales that bear pollen sacs in the male cones. Moreover we were to find, as it became possible to identify the fragile pollen of the yew, that it was present in large amounts in the shallow fens at the margins of the raised peat bogs in Somerset and elsewhere. Thus, although we cannot point to undisturbed fen woods with yew growing at the present day, the sum of evidence points strongly to the fact that this indeed is an entirely natural habitat for the tree. We may not very readily associate the evergreen trees with such situations but in the seminatural woodlands of temperate North America, the wet glades very commonly carry dark stands of the American arborvitae, that is remarkably yew-like in general appearance.

Where the succession of the vegetation was able to progress to the stage of a ground-cover of *Sphagnum* moss upon the leached and acidic peat surface, pine progressively came to replace oak, so that after a phase when both trees were present together, there followed an open woodland of pine upon the deepening mattress of saturated moss peat. This phase of the progress of natural vegetational change has certainly operated in many parts of the Fenland basin, moving most quickly where the soils were initially sandy and sterile and in areas subject to little inundation, but more slowly where the surrounding country was calcareous and flooding by alkaline water frequent. It is this process that produced the many layers of pine stools and fallen trunks encountered throughout the Fenland, not as part of the basal high forest, but within the peat at variously higher levels above it (Plate *12*). The buried layers of pine trees at Wood Fen, described first more than a hundred years ago proved to be a good instance of this category of buried forest and others have since been demonstrated along the east margin of the Fenland in the area of Woodwalton Fen, Ugg Mere and Glass Moor. An extraordinarily interesting parallel has since been literally brought to light in a buried forest on the south-western coast of Holland where dock excavation revealed the presence of a buried forest with stools and fallen trunks of over seven hundred trees *in situ*. Those rooted in sand at the base were both pine and oak, but above, embedded in the growing depth of acidic *Sphagnum* peat the trees were of even-aged pine growing with some birch on a forest floor strewn with pine cones and pine needles. Matching of the tree-rings of the trees, supported by radiocarbon dating of the innermost and outermost layers of several trees proved that the pines began growth within about a hundred years of one another, grew together for some two hundred years and finally all perished at much the same time without progeny, victims of the thickening moss layer. It is a story very applicable to our Fenlands, not least since the site lies in an analogous situation on the other side of the North Sea and was sealed in by the clay of a subsequent rise in sea-level.

The remains of the Fenland pines are particularly easy to recognise. In the field the flaking bark is characteristic, the wood still has its natural tough stringiness and red colour and may even keep its original smell of turpentine. A very characteristic system of natural decay of the sapwood often leaves the resinous heartwood of the centre of a pine stool with its radiating whorls of the bases of lateral branches. The needles can be seen alongside in the typical two-leaved short shoots, and very often there are mature cones in large numbers. Indeed needles and cones were so abundant at Wood Fen that it was hoped to use them to determine if the Fen pines might not have belonged to the native Caledonian variety *Pinus sylvestris* var. *scotica*, but the variability of the present-day Scottish material proved to overlap too greatly that of the continental and contemporary English stock. The task of recognising the development of the pine phase in the buried fen woods is made still easier by the extreme ease of identifying the pollen with its two rounded air-sacs (Fig. 5). Wherever timber of pine occurs we find the corresponding rise in frequency of this pollen type, often together with an increase of birch which alone among the deciduous trees can accompany pine on these wet acidic peats.

It is interesting that the birches should have this dual role in the progress of the fen wood succession, firstly as pioneers along with alder and finally with pine in the stage just prior to total burial of the fen woods below *Sphagnum* bog. Each phase of course is one of very open conditions that satisfy the birches' primary demand for light: the wide tolerance of soil reaction, nutritional status and wetness allow it to be present, mostly as the fen birch, in both the early and late stage. Although the timber decays very readily and requires microscopic examination for secure identification, the bark is generally well-preserved and generally recognisable though in the field that of hazel and alder is often mistaken for it. The winged fruits are abundant and easily recognisable in peat, occurring wherever the appropriate stage of fen wood development is found, and this is equally the case for the abundantly produced pollen, that may of course also come from birches growing on adjacent mineral soil. As we should expect from their rôle in the carr stage of the succession, willows were certainly part of the buried fen woodlands: their pollen is recognisable enough in the earlier and wetter stages and leaves of sallows are often encountered, but the timber, like that of the alder and birch decays very readily so that it is under-represented as trunks. The careful identifications by Dr Clifford of fen wood layers at various sites have disclosed such items as rose prickles, stones of purging buckthorn, wood and stones of dogwood, stones of blackberry, fruit and twigs of hazel, and these confirm our picture of what the undergrowth of the former fen woods was like although in some riverside sites components such as the hazel nuts and twigs may well have been water-borne and not part of the fen wood themselves.

5

Flandrian deposits and the Fenland Research Committee

Ten thousand years is a very short period indeed in geological time but the last ten millenia none the less have qualities and significance lacking in older periods. It was a time when European climate changed substantially, sea-levels altered, the earth's crust adjusted itself to relief from the burden of the northern ice sheets, coastal features changed correspondingly and rivers altered in volume, load and valley shape: the time was too short for extensive evolution of new species of plants and animals that might be used as zone fossils, the species are those of the present day but they have been subject to big migrations and the re-creation of communities destroyed by the last glaciation. The period is one through which we have arrived at the shape of our present-day landscape and only by having regard to the happenings of this latest geological period is the present condition of plant, animal and human cultures properly understood. It is deposits of this time, moreover, that lie nearest our hands since they constitute the most recent cover of the landscape. Our excavators and earth movers constantly work in Boulder Clays, outwash deposits, coastal dunes and marshes, estuarine silts and inland marsh and peat deposits that we encounter in our drainage cuts, roadworks and dock excavations and all of them offer us abundant evidence of conditions of the most recent geological past.

The move to make systematic enquiry into this period began early in this century in Scandinavia, and at the hands of a number of great pioneers, a co-ordinated pattern of events began to emerge that embraced the retreat stages of the Fennoscandian glaciers, the various changes of land- and sea-level (with drastic effects upon the Baltic Sea), climatic alteration, the building of lake and mire deposits, migration of fauna and flora and the succession of prehistoric man through from the Mesolithic hunter-fishers to Neolithic, Bronze Age, Iron Age and Viking man. Gradually the pattern of correlation tightened so that, for example, one could say that when the Baltic was a fresh-water lake discharging by a great river across south Sweden to the Atlantic coast the mixed oak–hazel woodlands were already being established and mesolithic man was living beside the Danish bogs and fishing its rivers. The Scandinavians also found it possible to furnish a time

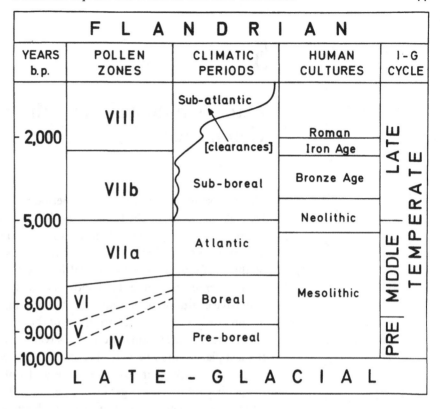

YEARS b.p.	POLLEN ZONES	CLIMATIC PERIODS	HUMAN CULTURES	I-G CYCLE
		Sub-atlantic		LATE
2,000	VIII	[clearances]	Roman	
			Iron Age	
	VIIb	Sub-boreal	Bronze Age	TEMPERATE
5,000			Neolithic	
	VIIa	Atlantic		MIDDLE
8,000	VI	Boreal	Mesolithic	
9,000	V	Pre-boreal		PRE
10,000	IV			

LATE - GLACIAL

Fig. 13. Table setting the most commonly used divisions of the Flandrian Period, the last 10 000 years, against the common scale determined by radiocarbon dating. All the divisions are approximate and not equally applicable over the whole of Britain. The extreme right shows the climatic divisions applied to all interglacial periods. The curve in the middle column is based on the pollen-analytic evidence for forest clearance in England and Wales at the hands of prehistoric and historic man.

scale from the counting of the matched series of annual 'varves', i.e., laminations in sediments of the series of lakes that had existed next to the retreating glaciers. We have already described how the methods of pollen analysis had been developed by von Post, methods so valuable for palaeoecology in their many geological applications that they have spread world wide and are still in intensive use, and have themselves great value in providing us with indirect geological time scales. The lesson was soon learned: it was evidently of great importance, when investigating any episode or object in deposits of this time, to place it securely in its stratigraphic position and to secure all possible evidence that would tie it within the general framework of Quaternary development (Fig. 13).

Whilst the earlier usage was to speak of this period since the glacial retreat from north west Europe as 'post Glacial', which in a loose sense it was, it was better geological practice to give it the name of a site at which its deposits had been thoroughly investigated, and it is accordingly now known as the 'Flandrian' from the coastal deposits of this period examined effectively in Flanders. It could not have been called Fenian for obvious reasons even had Quaternary research in the Fens progressed sufficiently when the term was adopted. The Flandrian had now assumed its place as the latest phase of the

Quaternary epoch, a period that covered two or three million years and took in many distinct glacial periods separated from one another by milder interglacial periods of which possibly the Flandrian is but the latest example.

There had been extremely little progress in knowledge of the geology of the Fenland basin since Skertchly had published his Memoir of the Geological Survey in 1877. Its scope, wisdom and far-sightedness were indeed against further publication although scattered finds of fossil animals or human artefacts had been reported now and then. The stimulus to fresh advance came, as I have already indicated, in the expansion of ecological ideas, particularly of dynamic succession, in the adoption of the techniques of pollen analysis, and finally in the whole concept of Quaternary research, which as Professor Grahame Clark explains, converted him to recognition that it was time that British archaeology developed from the stage of surface collection of finds that had 'not as a rule differed markedly in technique from garden digging', to a *stratigraphic* approach. In this there was concentration upon sites where archaeological material was found undisturbed in a sequence of geological deposits that with carefully exact excavation and proper examination might yield evidence both of date and environmental circumstances of the incorporation of the material and thus of the culture to which it belonged.

By the 1930s the time and circumstances were ripe for applying these new methods and concepts to the Flandrian deposits and prehistoric development of the Fenlands. Dr Cyril (later Sir Cyril) Fox had published his monumental *Archaeology of the Cambridge Region* recording a wealth of evidence from those fenlands and fen margins within his area. Moreover, his detailed distribution maps, especially those of the Neolithic and Bronze Age bore witness to past Fenland conditions, surprisingly favourable to occupation, and by contrast to extreme sparsity of occupation in Iron Age and Anglo-Saxon time. Dr Grahame Clark having just completed and published his survey of the Mesolithic age in Britain saw in the Fenlands promising opportunity to set the British cultures in their environmental context and provide them with the climatological time scale known already for the Scandinavian Mesolithic.

It was scarcely surprising that the Fenland Research Committee, a working group of both amateurs and professionals, established in June 1932 should have been an immediate success. It was under the wise and genial presidency of the then Professor of Botany in Cambridge, Dr A. C. (later Sir Albert) Seward, and its extremely active and acute secretary was Grahame Clark. The Vice-President was Major Gordon Fowler who had some time previously become manager of water transport for the beet sugar factory at Queen Adelaide, Ely. Fowler was a massive hearty man, with a career in

Canada and South Africa behind him and a distinguished war record that had cost him a leg: nevertheless he continued active interest in boxing, hockey and sailing and at Ely developed a very informed and scholarly interest in all the phenomena of the Fens, most especially in the many extinct natural waterways and undated drainage works. His work brought him into contact with fen farmers over a big area and he lost no opportunity of talking with them about the Fens or of visiting sites where they had unearthed things of interest. Later he stimulated interest by giving lectures at the Fenland villages. He was very widely known and few new discoveries failed to come to his notice. When a site he had visited seemed to merit closer attention telephone calls would inform the members of the committee most likely to be concerned and next day with gum-boots, spades and peat-indifferent clothing a small party would *rendezvous* at some agreed point, to be led to the site and there investigate, measure, photograph and sample it as seemed necessary. Aside altogether from his own research this liaison work made Fowler invaluable to the Fenland research and it was with great pleasure that we viewed the recognition of his work by the conferment of an honorary M.A. upon him by Cambridge University in 1948. Active at the side of Grahame Clark was C. W. Phillips an amiable, generous giant of a man whose gifts of total recall of all information read or heard and meticulous recollection of vast stretches of the English countryside were later to find very appropriate outlet as archaeological officer to the Ordnance Survey, and who meanwhile was to have the great pleasure and distinction of directing excavation of the Sutton Hoo treasure. Archaeology was served also by T. C. Lethbridge and Miles Burkitt, a link with the Abbé Breuil and the days of the earliest involvement of British archaeology with the geology of the Ice Age. A similar watchful and benevolent geological supervision was exercised by Professor O. T. Jones and no hypothesis of Fenland stratigraphic history was worth considering that would not withstand his experienced criticism expressed in a quiet but withering Welsh accent. As he was a member of my own college, night after night we sat together at dinner and a Fenland problem habitually lasted through coffee, and the trip home, finishing only in a cooling car in the road outside his front door. I was fortunate if an explanation I proffered was still intact after it had been through this mill! In his department at this time was also Dr W. A. Macfadyen an oil geologist who was practising the use of fossil foraminifera as geological indicators. They are marine organisms leaving delicate calcareous shells that can be sieved out of the containing silts or clays, and which under a low power binocular microscope can be referred to genera and species. They exist in vast numbers in the oceans still and display considerable sensitivity to the salinity of their environment, so that Macfadyen was able to deduce from their foraminiferal content the nature of the water in which the mineral sediments of the Fenland had been deposited. Geographical interests were well served by the consistent enthusiasm of Professor J. A. Steers and working in that department at this time, though independently, was H. C. Darby whose books *The Draining of the Fens* and *The Medieval Fenland* were to become classics of historical geography, and who was later to succeed Steers in the Cambridge Chair of Geography.

We owed much to the contributions of successive Chief Engineers of the River Great Ouse Catchment Board, notably Mr O. Borer and afterwards Mr W. E. Doran, O.B.E., not only because their close familiarity with the existing drainage system and ability to find highly skilled workmen facilitated our excavations, but because they were often able to provide us with useful stratigraphic data from existing records of borings made by their organisation and could often usefully criticise hypotheses about the former Fenland hazarded by those less closely in touch with the daunting realities of present-day conditions.

Many others came in to assist at the greater or less excavations of the committee, to attend its discussions or to contribute some special expertise. At its height it consisted of some forty-two members: many then or later occupied positions of considerable distinction and much helped the provision of grants, small by today's standards but essential then, for field operations and special equipment. We had the benefit, too, in the days when aerial photographic survey was in its infancy, of the benevolent interest of the then archaeological officer of the Ordnance Survey, O. G. S. Crawford, himself a pioneer in the application of air photography to archaeology. He not only initiated a Fenland air survey but loaned to the committee a set of large scale maps that we were able to employ as basis for cartographic record of our activities, whether separately published or not.

Of course major importance and interest attached to those special field projects conceived by the Committee for selected sites that were expected to yield high dividends in clear resolution of particular problems, in abundance of recovered material or in evidence of a long series of related events. Such were the investigations at Shippea Hill on the Littleport–Mildenhall road that are described in our next chapter. It was such enterprises, often spread through several seasons and involving many specialists, that provided the main threads of evidence and argument in establishing the continuity of past Fenland history. Although individual discoveries brought to notice haphazardly would involve separate quests on any stage of that history, many required attention then and there and were tackled with what scientific methods were then available and seemed appropriate. It lent special excitement that these problems arose so suddenly and were so individual, but we were always aware that in the end the solutions to these separate components of the puzzle must fit into one consistent picture, that of the historical evolution and structure of the whole Fenland and of its human, animal and plant occupants through past ages. This knowledge rested in the back of one's mind in trying to solve any one problem: its bearing on the sum of all the other problems had to be remembered, not always an easy matter standing in gumboots in a sloppy trench and trying to resolve procedure in dealing with remains in the oozing black wall of the trench dug through peat and clay.

After relatively few but very rewarding years of activity the members of the Fenland Research Committee were dispersed after 1940 about all manner of war-time activities and formal meetings were never resumed. Nevertheless the body had then served the function of having stimulated and focussed scientific interest from many directions upon the problems of Fenland history, and for that matter had generated the impulse to found similar research ventures independently in other regions. After the war also, as occasion permitted, some of the projects and sites opened by the committee were re-examined with the aid of new information and new techniques. More specifically pollen analysis and peat stratigraphic studies having been continued and extended in the Cambridge Botany School, the University set up there in 1948 a Sub-department of Quaternary Research whose Director was responsible to the three respective Professors of Botany, Geology and Archaeology, and Anthropology. It was in this sub-department that, with the help of money from the Nuffield Foundation, there was set up in 1952 one of this country's pioneer radiocarbon dating laboratories; in the hands of Dr E. H. Willis and later Dr R. A. Switsur the dates it provided were invaluable in checking the early conclusions of the Fenland Research Committee.

Already before the war-time dispersal however, the committee had been able to construct a comprehensive outline of Fenland history that had been elaborated in Transactions of the Royal Society for 1938 and given in condensed form in the guide book for the Cambridge meeting of the British Association in the same year. This outline schema had a great deal of internal consistency and happily the radiocarbon dates did not substantially modify it. This was the more convincing since many of the radiocarbon age-determinations were made upon the actual samples of peat or wood collected upon the original Fenland Research Committee investigations, whilst others had been obtained by revisiting or even re-excavating the original site.

In the accounts that follow it has seemed reasonable in each chapter to begin by reference to the chief Fenland Research Committee sites, but to add the results of radiocarbon dating and such further enquiries as have yielded important results. We follow, throughout the book, the now usual convention, of citing with lower case letters b.c., and a.d., all radiocarbon ages in years before or after the birth of Christ. Capitals are reserved by custom for actual (solar) years as known or deduced from other evidence. Differences between the two scales, apparent increasingly at ages greater than two thousand years are not such as to affect the broad trends of our argument. The continuing activities of the committee admirably reflect the central drive of human enquiry that arises from natural curiosity, and the scientific concern with causation and processes of operation.

6

Shippea Hill and the natural bed of the River Little Ouse

Shippea Hill is in the south-eastern Fenland, taking its name from an almost imperceptible rise in level where the Littleport–Mildenhall road meets the Fen margin. Excavations began there in 1931 with the investigation of a rich scatter of artefacts, flint, pottery and bones, on the surface of a low sandy ridge just projecting above the general cover of black fen peat, near Plantation Farm (Fig. 14). The site lay inside the deep loop of a raised silt bank or 'roddon' that Gordon Fowler had shewn to be the natural, but long-abandoned, channel of the River Little Ouse. Careful excavation down the flanks of the sand hill soon proved that the artefacts, the bulk of which were relatable to the Early Bronze Age, stratified outwards into the peat on its northern side forming a layer a few inches above its lower surface where it sat on the sticky 'blue buttery clay' that we rightly took to be part of the Fen Clay that Skertchly had shewn to be a major feature through the peat fens, separating the Upper Peat from the Lower. A line of hand-borings from the

Fig. 14. The natural course of the River Little Ouse. The more important natural river courses are shown by continuous black lines: where still carrying active streams they are reinforced by broken lines. Artificial channels are shown by broken lines, the most recent of them the marginal cut-off channel opened in 1964. Note the large shell marl area of Red Mere formed in apposition to the Little Ouse roddon. Several sites important in elucidation of Fenland history are shown, Shippea Hill, Wood Fen, Southery and the Romano-British fen-margin settlement at Hockwold-cum-Wilton (H-c-W).

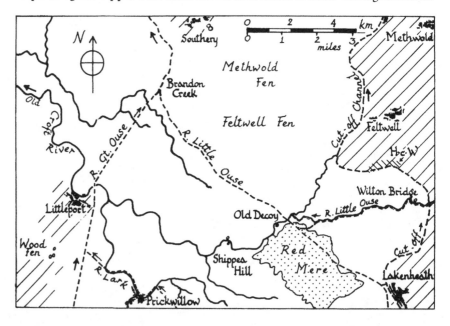

Plate *15*. Shells of fossil foraminifera recovered by washing and sieving from the Fen Clay at Ugg Mere, Hunts., from the Green Dyke section (see text). They are about the size of a pin head but under magnification exhibit microscopic features that allow easy recognition of the various species and genera. Dr W. A. Macfadyen identified this group as species that live in brackish water with salinity much below that of sea water, especially (*a*) *Nonion depressulus*, (*b*) *Elphidium excavatum* and (*c*) *Rotalia beccari*: the fourth species shown, (*d*) *Lagena clavata*, has a wider salinity range. (Photograph by D. Newling.)

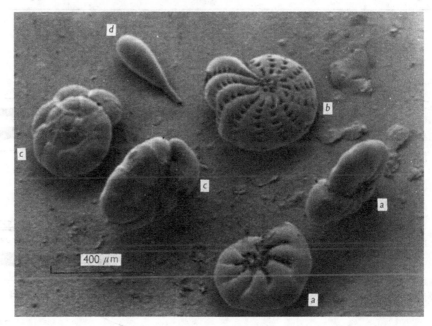

excavation towards the roddon disclosed some 5 to 7 ft (2 m) thickness of Fen Clay overlying a similar depth of the Lower Peat, and it was evident that the sand hill was indeed upon the bank of a former deep river channel whose width much exceeded that of the roddon itself. Outside the sand hill the fen floor was much shallower. That the deep channel was of considerable age was shewn by pollen analyses which proved at each bore hole that the base of the Lower Peat had formed in the Boreal period, when pine was a dominant forest tree and hazel was still very abundant, with an evident transition to the Atlantic period recognisable some 2 ft higher where alder first became outstandingly abundant. What was more, while examining the contents of the drill from the base of our deepest bore, I recall picking out what I took to be a bit of pine bark: licking it clean of peat (a standard and effective field procedure for seeds and such) to my astonishment it shewed itself to be a struck flake of flint soon confirmed by Grahame Clark as like those of Mesolithic cultures. Since there was a small but absolutely definite microlithic flint component of the same period in the surface litter of the sand hill, we could glimpse the exciting possibility of tracing this culture also laterally down into the deep channel deposits.

Meanwhile we continued investigation by seeking to tie into the sequence of channel deposits those of the latest phase, the meandering raised roddon. The silts of which it is mostly composed defeated our Swedish hand auger and it was necessary to employ professional drillers using casing. Their records proved the silts to a maximum depth of about 17.5 ft (5.3 m) below

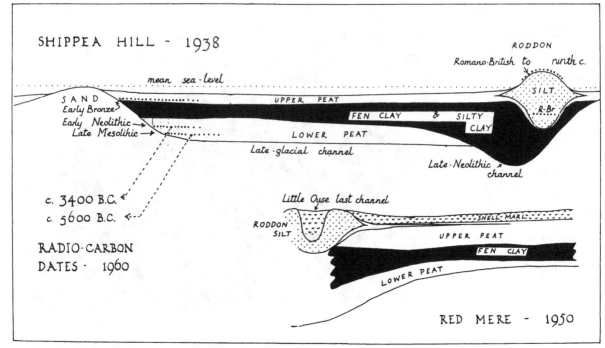

Fig. 15. Diagrammatic section of deposits in the natural channel of the River Little Ouse at Shippea Hill. The sandy river banks were occupied by prehistoric man who left his artefacts stratified into the channel deposits; these and the radiocarbon dates suffice to date the sequence of events. The latest phase so defined is that of the roddon of tidal silt (Romano-British, R-Br) and the associated shell marl of the freshwater Red Mere.

the top of the roddon ridge now itself standing about 5 ft (1.5 m) above present mean sea-level: they were contained in a peat-lined channel which was continuous with the Upper Peat and certainly related to some stage of it. The fossil foraminifera clearly shewed the roddon silts to have been laid down in tidal estuarine conditions, i.e., in distinctly more saline water than that from which the Fen Clay had been deposited. The foraminifera of this earlier bed were laid down in brackish water suggestive of lagoon conditions rather than those of a channel, lagoons subject to tidal incursions now and then bringing in indicators of open salt sea and derived fossils from the far older Chalk of the coast. The Fen Clay itself at its deepest went down to some 28 ft (8.5 m) below sea-level, having occupied a channel cutting through the lower peat that here reached about −22 ft (6.7 m). It will be seen that substantial past changes in relative land and sea level were indicated, and it was of interest that already in 1931/32 we were aware of the probable Romano-British age of the roddon silts, since a pot of that age had been recovered a few feet below the surface whilst the foundations for a building on the roddon crest nearby were being dug (Fig. 15). In the months following the excavation the committee heard reports upon the non-marine shells, the bones of domestic and wild animals, plant remains and pollen analyses. These encouraged the opening of a second excavation during the summer of 1933 on the opposite side of the primary channel and of the roddon, at a similar sand-ridge no more than 350 yards (320 m) from the earlier site. This we called the 'Peacock's Farm' site and again it aimed to trace down into the channel deposits the rich scatter of archaeological objects on the surface of the sand hill, most particularly those of Mesolithic age that were found fairly abundantly on this hillock. Our excavation had to go to a depth of at least 16 ft (5 m) in beds liable to collapse and severe flooding had to be expected. With the help of very expert fenmen, adequate pumps and careful progression a series of pits of increasing depth was excavated, the lowest reaching 19 ft (5.8 m) below sea-level (Plate 16). Again the Early Bronze Age artefacts extended into the base of the Upper Peat but it left a surface residue to be followed deeper. The Mesolithic flints were shewn to rest in the Lower Peat 4 ft (1.2 m) from its upper surface at an horizon defined by the pollen analyses as late Boreal, and where the sandiness of the channel peat reflected the occupation and disturbance of the vegetation cover of the sandy river banks. More unexpectedly, 2 ft (0.6 m) above this level there was found in the Lower Peat a sherd of 'Windmill Hill' or Neolithic 'A' pottery, other samples of which formed part of the surface scatter descending the flanks of the channel. We were not at this time able to

Plate 16. Completed excavation made in 1934 through the channel deposits of the natural Little Ouse river channel at Shippea Hill. Professor Clark examining the early Bronze Age level, Gordon Fowler standing on the sandy bank of the channel, on which were to be found the mixed artefacts of the three cultures that were stratified into the channel beds as shewn in the key.

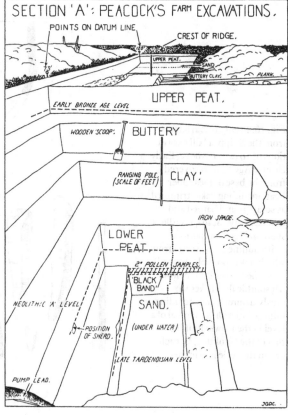

SECTION 'A': PEACOCK'S FARM EXCAVATIONS.

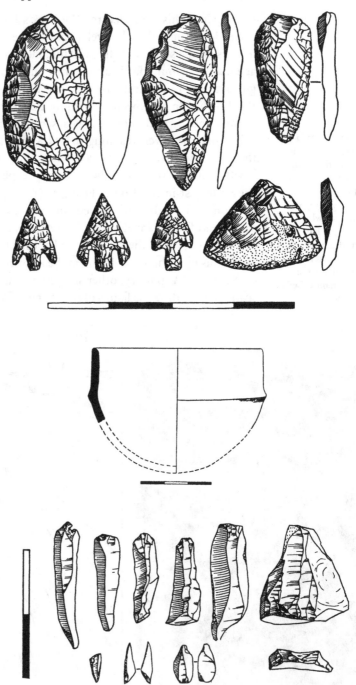

Fig. 16. Prehistoric artefacts from the Shippea Hill excavations of 1932–34. Above, Early Bronze Age flints recovered from the base of the Upper Peat, including flakes trimmed as scrapers and barbed and tanged arrow-heads; middle, a sherd of Neolithic pottery (with outline of the vessel from which it was possibly derived) from the Lower Peat; below, Mesolithic flint flakes variously trimmed and including minute 'microliths' typical of a level in the Lower Peat below that of the Neolithic. In each group the scale is in inches.

characterise the Neolithic horizon in pollen-analytic terms except to refer it generally to the Atlantic period, but fresh-water mollusca from somewhat above it in the Lower Peat were shewn to contain an abundance of one species with a very southern distribution in Europe and now rare in Britain: it suggested a climate probably warmer than that of today.

What our two excavations had now achieved was the recognition of no less than four archaeological levels in stratigraphic sequence and firmly related to their contemporary environment and stage of Flandrian history. Respectively they were the late-Boreal Mesolithic deep in the Lower Peat of the channel and below present sea-level, the Atlantic Neolithic 'A', somewhat higher in the Lower Peat, the Early Bronze age, after the intervention of a phase of brackish water invasion, in the lower layers of the Upper Peat, and the Romano-British of the roddon silts formed in a later phase of marine incursion into the river channel. These two exercises by the Fenland Research Committee have properly been regarded as something of a landmark in British archaeological method, convincingly demonstrating, as they did, the enormous potentialities of a stratigraphic approach in sites that have been carefully selected for their Quaternary geological setting: it heralded a widespread change in manner of approach to the problems of British prehistoric archaeology. The many ancillary results that stemmed from the Shippea Hill excavations of 1931–33 will most conveniently be mentioned in the following accounts of successive stages of Fenland's Flandrian history. They had of course also shewn to individual research workers of the committee the many still unsolved problems of the history of the natural Little Ouse river and at intervals specific fresh investigations were undertaken. In an attempt to see what the channel deposits were like nearer the upland, a few borings were made in 1935 at Wilton Bridge some 5 miles (8 km) upstream from Shippea Hill. These proved the original river here to have been beyond the limits both of the Fen Clay and of the roddon silts, but the continuous peat extending down to 7 ft (2.1 m) below sea-level was shewn by pollen analysis to be of Boreal age like that of the Lower Peat at Shippea Hill. It had, however, in its upper layers indications of forest history later than anything then seen in the Fenland with much pine, birch and even beech, the latter presumably reflecting local woodlands of that tree growing on the local chalk exposures. Alongside the Little Ouse roddon just above Shippea Hill there had long been known a very large area of white calcareous marl, presumably the residue of a former fresh-water lake, but with no associated discoveries from which its date might be guessed. Stimulated by local lectures given by Gordon Fowler about the Fenland, an imaginative Ely schoolboy, Anthony Vine, undertook to trace and map on the six-inch survey maps the limits of the lake as disclosed by the extent of the shell-marl in all the drainage ditches of the area. He was strikingly

Fig. 17. Tree-pollen diagram from samples taken at the deep excavation in the natural channel of the River Little Ouse, 1960, with radiocarbon dates inserted at their sampling levels (by the convention now used these dates should be 'b.c.'). The pollen analyses correspond closely with those obtained in 1935 from boring at an adjacent point in the channel: the dramatic rise in frequency of lime, alder and ash (*Fraxinus*) indicates the boundary between pollen-zones VI (Boreal) and VII (Atlantic) (see Fig. 18).

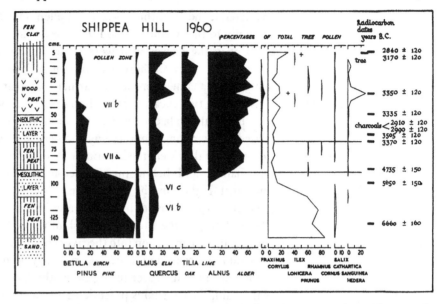

successful and the mapped outline of the lake that he produced bore such evident and close relationship to the roddon along its north-western flank that it roused the strongest suggestion of contemporaneity and possibly of cause. It was this suggestion that was taken up by J. N. Jennings, now Professor of Geomorphology in Canberra, during field and laboratory studies reported on separately in Chapter 9.

By 1960 the Cambridge Radiocarbon Dating Laboratory was in full operation and Grahame Clark and the author undertook to reopen an exposure of the channel as close as practicable to the original excavation at Peacock's Farm, so as to date the leading horizons of the Lower Peat and its contained cultures by this new physical means of fixing absolute (or nearly absolute) ages in years (Fig. 17). At the same time advantage was taken of the opportunity to repeat pollen analyses through the Lower Peat at closer intervals than before and taking advantage of the greatly improved standards of microscopic identification and interpretation over the intervening twenty-five or more years. These results are best considered in the next chapter on the Lower Peat of the whole region, but it may at once be said that this work was done at a time when the absolute age of the Neolithic in the British Isles was being subject to a quite unexpected revision downwards, and that these Shippea Hill results fitted a substantial scatter of radiocarbon dates elsewhere in placing the opening Neolithic at least as early as 3000 B.C., thereby also trebling the range of time hitherto ascribed to its duration and removing the established conjecture that in this country the culture had recognisably spread outwards from Wessex and Sussex.

The base of the Lower Peat dated as late Boreal at both the Shippea Hill sites indicated the infilling where waterlogged conditions prevailed across the floor of the channel, but that very wide bed, far too wide for the modern river, must have been cut when it carried a vastly greater volume of water, such as might have accompanied the general thaw at the end of the last glacial period. It seemed possible, therefore, that the channel somewhere contained peats or organic muds extending back in time towards that event. They were found by boring in 1937 at 'Old Decoy' a site some 2 miles (3.2 km) upstream from Shippea Hill at a point guessed to be in mid-channel. The Lower Peat here underlaid 9 ft (2.7 m) of Fen Clay and itself reached a thickness of 6.5 ft (2 m). A close series of pollen-analytic samples shewed that it extended through the early Atlantic as at Shippea Hill, but below embraced not only the whole of the Boreal period with its characteristic sub-zones, but below this the Pre-boreal period which is the opening zone of the Flandrian, already known as commencing about 8000 B.C. The greater age of this lowest foot or so of peat was clearly shewn by a pollen composition strikingly different from that of deposits above. Pollen of herbaceous plants quite outnumbered that of trees, and pointed to a generally open vegetation in which birch and willow were the only local trees and were far from forming continuous woodland. This is a type of vegetational cover highly characteristic of the late-Glacial in lowland England, and our attribution was strongly confirmed when Miss Robin Andrew, who had done the original pollen counts, returned to her prepared microscopic slides some years afterwards when improved standards of identification now allowed her to recognise from these levels, grains of many herbaceous plants common in southern Britain at that time but subsequently rare or extinct there, of dwarf birch and crowberry, and along with many aquatic and marsh plants, grains of many herbaceous plants characteristic of this period, such as *Artemisia* (wormwood), *Rumex* (docks), *Botrychium* (moonwort), *Lycopodium* (clubmoss), *Thalictrum* (meadow rue), *Helianthemum* (rock rose) and *Plantago* (plantain), with very abundant grasses and sedges. There was even one grain of the rare steppe switch-plant, *Ephedra*. We were afterwards to find pollen-analytic evidence of a similar kind from meres and channel deposits in the Breckland just at the edge of the catchment area of the River Little Ouse itself (see Fig. 18). Such assemblages became extinct or restricted to high altitudes with the subsequent spread of closed forest.

With this age determination we were brought much closer to explanation of the size and depth below present sea-level of the old channel. It took us back to the period of the great eustatic fall of ocean level during the last ice age described already (Chapter 2) when so much water was locked in the world's ice sheets that the North Sea had its coast line beyond the Dogger Bank and all the great rivers of north-western Europe were cutting their

Fig. 18. Pollen diagram from a deep Breckland basin near the eastern Fen margin. It covers the end of the last glacial stage in zones I to III, through which shrub and herb pollen is extremely abundant and the only common tree pollens are those of birch and pine: deposits of this character occur also in the bottom of the Little Ouse channel close by Shippea Hill at Old Decoy (see text). All but the most recent part of the Flandrian is embraced by the rest of the diagram that is divided into the quasi-chronological zones IV to VIIb commonly in use for England and Wales. They are all tree-dominated until the forest clearances that began early in zone VIIb.

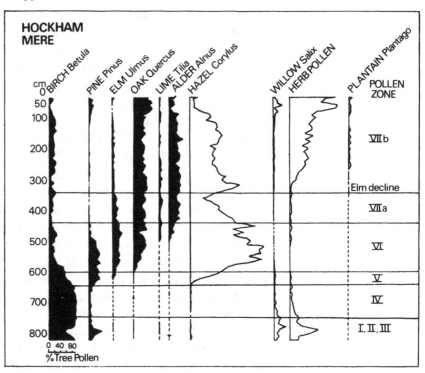

beds down towards base levels vastly below present sea-level. No doubt their volume and burden of erosive mineral matter were greater during the terminal phase of thaw. Such deep Glacial and late-Glacial channels have long been known for our British rivers and the Little Ouse could of course not fail to shew it since upstream as far as Cambridge even the River Cam runs above just such an impressive ancient channel. To the pollen evidence of the moorlog that the North Sea was still *terra firma* in the Boreal period, it had been added that Mesolithic man was present at a site now more than 100 ft (30 m) below sea-level, and it is interesting to have found artefacts of the same broad cultural category stratified into the Shippea Hill deposits also. To judge from analyses made in Swansea Bay, the general recovery in ocean level was now reaching its last stages although brackish water was not yet evident in the Fens. Up to this time the Fenland peat deposits were almost entirely confined to such deep wide valleys as that at Shippea Hill cut by the meltwaters at the end of the Ice Age. These now turned into fens through which meandered a diminished stream. Outside these few parts of the present Fenland, there were few areas sufficiently waterlogged to form peat, except perhaps local depressions in the Boulder Clay. The widespread extension of peat formation everywhere awaited some cause powerful enough to cause widespread and sustained high water-levels: such might have been assisted by the accepted general turn of climate to increased wetness at the opening of the Atlantic period, but it can scarcely have been the prime cause. This must almost certainly have been the continuing rise in sea-level which, as it brought brackish and even salt water into the coastal part of the Fenland basin, prevented discharge of the rivers and caused a backing up and general rise of fresh water in all the landward area of the Fen basin.

7

The Lower Peat and the Fen Clay

The Fenland Research Committee were already aware of a good deal of incidental evidence of the time at which the Fen Clay marine incursion had induced widespread general formation of the Lower Peat and brought about the widespread destruction of the towering high forest that up to that time had clothed all the Fen basin. Some of this evidence has been mentioned in describing the buried forests, but this mainly consisted in records of polished stone axes and bronze objects now in museums or collections with such vague attributions as 'Burwell Fen' and no indication of depth or enclosing deposit. Many such objects were found during coprolite digging where some may have lain on the forest floor whilst others came from an unspecified depth in the peat above it, so that they afford only a very general idea of the age of the commencement of peat growth. The case is somewhat better for a remarkable discovery now displayed in the Sedgwick Museum of Geology in Cambridge. It is a complete skeleton of the extinct wild ox, *Bos primigenius*, a massive animal with very wide spread of horns that was found with a triangular hole smashed through the frontal bone of its skull. Beside it in the peat lay a polished stone axe which was picked up and shewn to fit exactly into the fracture. Some Neolithic man of no mean courage must have driven his hafted axe into the beast's forehead, and one supposes that the stricken animal escaped his hunters to collapse and die in some area wet enough for the skeleton to have been preserved (Plate *17*).

Vague though these indications of age were, we also knew the level of peat growth in the Shippea Hill channel at the opening of the Neolithic (later shewn by radiocarbon dates to be about 3000 b.c.) and from the level of Fen floor outside the Little Ouse channel it was apparent that general flooding would have been not very much later than this. So much was indeed confirmed when it became possible to give radiocarbon dates to trees of the high forest killed and entombed by the general waterlogging. Samples from the wood of five separate large oaks from Isleham Fen, Mildenhall Fen, Adventurers' Fen, Wicken (2) and Adelaide Bridge, Ely were respectively 2051±66 b.c., 2254±60 b.c., 2430±140 b.c., 2655±110 b.c., and 2545±120 b.c. Since we now have a large body of radiocarbon dates

Plate *17*. Skull of the aurochs (*Bos primigenius*) from Burwell Fen, Cambridgeshire. The polished stone axe was found immediately alongside and evidently fits the hole in the frontal bone where it has been photographed. The extinct ox skull almost certainly came from the base of the peat at the level of the buried high forest, as the Neolithic axe suggests.

through the British Isles, placing the Neolithic period between about 3500 and 2000 b.c., it is evident that the buried high forest and its enclosing peat in the margins of the South Level are indeed of this time. This fits uncommonly well with the discovery made in 1938 during widening of the Swaffham Engine drain near Upware. It was a polished Neolithic stone axe that was found to have been made of a recognisable hard crystalline rock, used by an axe factory at Graig-Lwyd on the slopes of the Welsh mountain, Penmaenmawr (Fig. 19*a*). We were able to recover the horizon in the peat in which the axe was found, and although this site is just beyond the limits of Fen Clay deposition, it was possible to shew from pollen analyses and the

Fig. 19. Prehistoric axes from Fenland deposits, all related to the stratigraphy and the vegetational history. *a*, polished Neolithic stone axe from Swaffham Engine drain, Upware; *b*, Middle Bronze Age axe from Castle Hills Farm, Woodwalton; *c* & *d*, two Late Bronze Age socketed axes from the Stuntney bronze-founder's hoard.

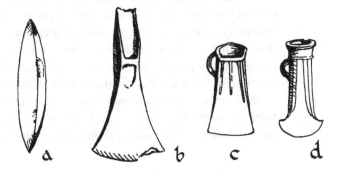

larger plant fossils composing the peat, that the axe was found just where fen conditions were at their wettest, i.e., the maximum of the Fen Clay transgression.

As the general waterlogging reflected the last stages of the progression of the marine transgression to its maximum what we clearly most needed to establish was the age of the peat just below the Fen Clay, especially where the clay reached its greatest extent as evidenced by its tapering out into the continuous deep peats never subject to salt water. This became possible when we could return to secure radiocarbon dates for sections and sites discovered by the Fenland Research Committee over twenty years previously: others were at sites found in the interim. At Wood Fen, the very edge of the Fen Clay was dated 2245 ± 110 b.c., and pinewood flooded by the salt water at Glass Moor in the Middle Level 2395 ± 110 b.c. Further seaward, at St German's, where the onset of the marine transgression effects was naturally earlier, the peat immediately below the clay bed corresponding to the Fen Clay yielded a date of 2740 ± 120 b.c. Sparse though the data are, they suggest that general waterlogging may have preceded the actual laying down of the brackish-water clay in a given site by three or four hundred years, a lapse that would indeed have allowed for the sequence of vegetational change recorded in the deposits at Wood Fen, time for killing the basal high forest and growth of a generation or two of later trees.

The Fen Clay that forms the middle layer in the sandwich of black fresh-water peat over so much of the Fenland is characteristically an unctuous soft clay of high water content and well described in the Fen term as 'blue buttery clay'. It owes its colour to the reduced state of the iron compounds in it and where drainage has exposed its upper surface, or the tubular stems of reeds have admitted oxygen to it in their proximity, it has always become orange brown. The brackish conditions of its deposition were decisively proved by Macfadyen's analyses of the contained foraminifera at Shippea Hill, but from wherever else samples of Fen Clay were submitted to him similar results were obtained. The conditions of lagoons intermittently invaded by tidal water of course provided just the quiet water needed for the settling out of the uniformly fine clay particles that give the clay its 'buttery' quality. This, however, is not to say that this texture was really constant: nearer the sea and in the channels by which the tidal water moved, that is, when currents were more rapid, the deposit was always siltier though clearly part of the same bed. The Fenland Research Committee was at one stage invited to believe that the Fen Clay originated in flood deposits washed down from the uplands during the wet Atlantic period, but had this been so, in tracing the clay upstream towards its source it should have become coarser and thicker, at first becoming largely silt, then sand and finally river gravel. The boring at Wilton Bridge, upstream in the River Little Ouse channel from Shippea Hill and Old Decoy, proved that, on the contrary, towards the upland margin the Fen Clay disappeared.

The essentially semi-marine origin of the Fen Clay is strongly attested by a great deal of other evidence. When the equivalent clay layer at St German's was examined it proved to have not only the expected foraminifera in it, but also shells of the cockle, *Cardium edule*, and, as we saw in Chapter 3, the pollen analyses of the uppermost and lowest layers shewed substantial amounts of pollen typical of salt-marsh vegetation. The same proved true, though less strongly so, at the various inland sites where pollen series passed through the contact zones. Furthermore, whenever the Fen Clay is dug through the contact layers they are seen to be grown through by the abundant vertical stems of the giant reed *Phragmites*, that tolerates brackish water readily and over flooded salt marshes on the Norfolk coast will form dense and tall communities, such as also form reed-swamp on the shores of the Baltic. The Fen deposit is a yellowish green, full of *Phragmites* stems and rootlets, with such a foetid odour that the fenmen call it 'bear's muck' or 'devil's dung', names that have their close equivalents on the marshes of north-west Germany and Holland where such deposits are '*Darg*' or '*Hundeköd*'. Still smaller fossils than the foraminifera are the fine two-valved siliceous shells of the algal diatoms, the marine forms of which are typically circular whereas the fresh-water ones are ovate or boat-shaped in normal aspect. When the same Fen Clay samples were examined both for diatoms and for foraminifera, as at Ugg Mere, precisely similar conclusions were independently drawn from each. At the other end of the scale we have the records of an occasional grampus or even of whale, a skeleton of which was found in the Fen Clay as far inland as Ely, whither the animal had doubtless penetrated by the main tidal channel.

It seems evident from both foraminifera and diatoms that the base of the Fen Clay was the most saline stage, and at Glass Moor Dr Clifford found the valves of the cockle in position of growth on the pine trees that had been covered by the invasion of tidal water: they also grew on the surface of the Lower Peat in the bed of Whittlesey Mere and accordingly lie now at its junction with the Fen Clay. This early maximum saltiness seems to tie in with the fact that in the old river channels especially, the Fen Clay occupies channels cut deeply through all the preformed Lower Peat and even into the fen floor. The most likely explanation is that gradual subsidence led finally to a sudden break-through of tidal water into the existing fresh-water channels and that it was the scour of tidal movements both into and out of the big newly flooded tracts that permitted the local downcutting.

As the Upper Peat now wastes away over very large areas of the Fenland, the upper surface of the Fen Clay is being revealed and it shews, especially to air-photography, its whole extent covered with a densely-branching pattern of shallow ridges (see Plate 23). This is now susceptible to mapping on quite a fine scale, as appears in the elegant coloured soil map of the Ely region published in 1974 by the Soil Survey. They suggest, certainly correctly, that this ramification is the remains of the creek system that drained the great salt-marsh which was the final phase of the Fen Clay deposition. It is a conclusion very strongly supported by the detailed dyke-section at Ugg Mere published in 1938 (Fig. 20) that exhibits such channels both cutting down into the Lower Peat and raised as low ridges just perceptible at that time as banks projecting through the cultivated Upper Peat. Pollen of salt-marsh plants and both diatom and foraminiferal analyses showed brackish water conditions and prompted speculation that the ridges were salt-marsh creeks, a view supported by the fact that some of them showed the central channel between the two raised banks constituting each ridge. It seems from the air photographs that the main channels along which the Fen Clay entered the branching system of finer creeks were the original deeply cut natural river beds of the Fens but this remains to be confirmed in convenient sections or by deliberate excavation.

Although the Woodwalton, Ugg Mere region of the Fenland clearly displayed a similar threefold stratigraphy of Lower Peat–Fen Clay–Upper Peat to that of the South Level where the Fenland Research Committee's work was mostly centred, it was evidently desirable to consider whether the Fen Clay was indeed the same formation in both areas, and Dr Clifford and I tackled this in the mid 1930s. From the deep borings of the Drainage Board along converging lines towards Wisbech it was possible to conclude that the Lower Peat and Fen Clay extended uninterruptedly seaward at least as far as Guyhirn, less than 5 miles (8 km) south-west of Wisbech. We

Fig. 20. Section along Green Dyke, near Woodwalton, showing the relation of the Fen Clay to the underlying Lower Peat with its numerous buried trees, and to the Upper Peat. The Fen Clay fills channels cut down into the peat below it and makes corresponding ridges projecting into that above: these are the traces of the salt-marsh creeks active during the clay deposition. The scale is greatly exaggerated vertically: the ridges are only about 1 ft (30 cm) high and 20 to 100 ft (6–30 m) across.

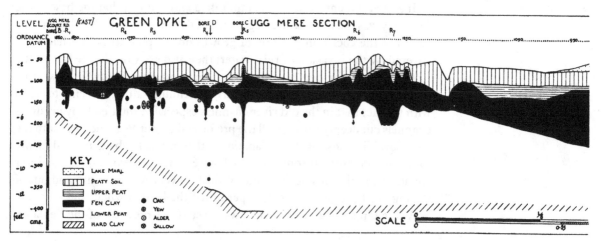

Fig. 21. Jet beads and bronze pin found upon 'Nancy', the Early Bronze Age skeleton recovered from Southery Fen. Her level in the Fen beds showed that the Fen Clay deposition probably ended before the opening of the Bronze Age.

proceeded from there to establish a line of levelled borings eastward round the gravel island of March and then south-eastward to finish up at Wood Fen, Littleport, where we were already satisfied of the general sequence of beds throughout the South Level. It was a distance of about 15 miles and over much of this distance our survey involved penetrating not only the Upper Peat but also the later silts that cover it on the seaward side of the Fens, and each boring had to be related to the nearest trustworthy bench mark. Though laborious, this task decisively shewed the continuity of the Fenland strata, the sequence, character and levels being consistent throughout, the upper surface of the Fen Clay varying only a little in height from mean sea-level to about 4 or 5 ft (1.5 m) lower.

We have already seen that there is no conclusive evidence that the Lower Peat anywhere contains remains of a culture later than the Neolithic, a conclusion reinforced by the radiocarbon dates for its uppermost layers. As might be expected from the nature of the Fen Clay, it has itself yielded no good record of archaeological objects found within it. We know, however, from both the Plantation Farm and Peacock Farm sites, that at Shippea Hill early Bronze Age remains were stratified into the base of the Upper Peat. This stratigraphic correlation was dramatically reinforced for us quite early in our research. In the Fen close to the island of Southery a female skeleton (referred to in the Fenland Research Committee's discussions by code-name 'Nancy') was found in the bottom layer of the peat only 3 in above the surface of the Fen Clay. 'Nancy' was referable to the early Bronze Age because she was still wearing a bronze pin and a necklace of the biconical jet beads that are characteristic of this culture (Fig. 21). Pollen analyses of the base of the Upper Peat here were closely like those at Shippea Hill. It is a pity that more precision does not attach to the ancient record of the discovery in the fens near Chatteris of a wooden boat sitting on the clay below the peat and containing a bronze rapier.

However, here again radiocarbon dating appears to have the last word. Its results indicate the onset of the Fen Clay in the landward Fens to have been generally about 2600 b.c., though a little earlier seaward and in the deep river channels. The top of the transgression at Wood Fen was at least 2245 ± 110 b.c., and two dates for the bottom of the Upper Peat at Denver Sluice were respectively 2440 ± 120 and 2135 ± 110 b.c. Nearer the sea at Saddlebow, only 1.25 miles (1.6 km) from St German's, samples from just above grey-blue Fen Clay, with *Cardium edule* (cockles) abundant in the top 6 in (13 cm), were dated 1945 ± 120 and 1950 ± 120 b.c. This sorts well enough with the date generally taken for the beginning of the Bronze Age in southern England around 1800 B.C. At most, therefore, the Fen Clay transgression occupied about a thousand years and in the landward Fens, where only its maximum was experienced, perhaps only half of that. The marine transgression was surprisingly short considering how great were the effects it produced, but the time scale agrees well enough with that of comparable effects in the Netherlands.

It is of interest before concluding this chapter, to consider the evidence jointly furnished by the Lower Peat and the Fen Clay for the climatic conditions of the time. The period in which they formed, the Atlantic, is normally held to embrace what used to be called the 'climatic optimum' no doubt better named the thermal maximum, of the Flandrian period, long associated by the Scandinavians with mean summer temperatures some 2° or 3°C higher than those of today. The British pollen diagrams, like their own, point to the greatest luxuriance of deciduous high forest in this period and in England the lindens, most warmth-requiring of the forest trees, reached their highest pollen frequencies during this time. The bulk of the linden pollen we can now say came from the winter lime, *Tilia cordata*, but within this time span there was also present the more demanding summer lime, *Tilia platyphyllos*. Both species were subsequently much reduced in frequency, partly no doubt by economic exploitation of their bast fibre as well as the effect of climatic 'deterioration'. It was of very special interest, therefore, that in the resumed Peacock's Farm excavation of 1961 we actually recovered the unmistakeable flower of *Tilia platyphyllos* from the Neolithic horizon, clear evidence of the local growth of this tree until recently not even accepted as a British native. From the shallow water terminal phase of the Fen Clay at Peacock's Farm among the fresh-water mollusca attention had been drawn particularly to *Pseudamnicola confusa* a 'decidedly southern species' now very rare in the British Isles and to be regarded as a relict of more congenial times.

From the buried oak at Isleham Fen whose radiocarbon age has already been given, there was found in the sap wood a wood-boring insect, *Cerambyx cerdo* not now known in Britain and indeed confined in Europe to the south and south east. It had also been found fossil in buried Danish oak forest and likewise employed as a thermal indicator.

Among birds the two European species of pelican are also distinctly southern in their present day range, which does not extend to Britain. Nevertheless it was at Saddlebow in the bottom layers of the upper peat, whose radiocarbon age has just been given as about 1950 b.c., that a bone of the genus *Pelicanus* was collected by the chief engineer in charge of the excavation of the flood relief channel. No fossil records of this bird are known in Britain later than the pre-Roman Iron Age lake-village at Glastonbury.

It would be pleasant if one could add a fossil reptile to this list of thermal indicators but the remains of pond-tortoise cited as found in Feltwell Fen are elsewhere referred to Wretham Mere in the adjacent Breckland. This again is a fossil found in Danish peat deposits well outside the present range of its breeding populations that now lie substantially further south. The pond-tortoise buries its eggs in warm dry waterside banks to hatch in the heat of the summer and certainly the sandy Breckland soils, which surround Wretham Mere and also stretch below parts of Feltwell Fen, would have been suitable to it. We have, however, yet to find its highly recognisable carapace stratified into datable deposits. It is from another of the former Breckland lakes, Hockham Mere, that we have very clear records of the fruits of a thermophilous submerged water plant *Naias marina*, abundantly at the transition from the late Boreal to the early Atlantic period. Thus the fossil Fenland records for warmth-requiring organisms fit easily into a much wider pattern of evidence, not only in this country but throughout north-western Europe.

8

The Upper Peat: hoards and trackways

When the Fen Clay marine incursion had reached its greatest extent and the lagoon clays had extended to within very short distance of the surrounding uplands, a phase of constancy, or more likely of regression of sea-level led to the widespread formation of fresh-water peat above the Fen Clay surface. The 'Upper Peat' formed in this way is a bed of substantial thickness traceable continuously in sections and borings from the Fen Clay edge seaward, so that it forms a consistent feature of the big drainage cuts such as Popham's Eau (see Fig. 37) and the new marginal relief channel between Denver and Lynn, and at such sites it is apparent that it extends still further towards the Wash beneath the silts of the later marine transgression. It was evident in the St German's excavations as the 'two-foot peat bed' where the local pollen sequence revealed the transformation of the Fen Clay surface from salt-marsh, through fresh-water fen and through alder fen wood to fen wood with oak trees. For a site so far seaward to exhibit such a succession implies a sustained freedom from flooding and the evidence of sites further inland strongly confirms this. For a time the flat surface of the Fen Clay was flooded with fresh water despite erosion channels, now recognisable as 'old runs' cut into its surface here and there as at the Southery site, where 'Nancy' was found. At Shippea Hill and Ugg mere is clear evidence from the abundant fruits, stems and pollen of aquatic plants that there was a short phase of shallow open water, but very soon this was invaded by sedge-fen and then by alder fen woods. How far the vegetational succession progressed naturally depended greatly on the local conditions. However, as early in the life of the Fenland Research Committee as 1933 we were astonished to find at Nordelph, some 2 miles east of the Fen margin at Denver, that it had progressed as far as the stage of acidic raised bog. In the middle of the Upper Peat, there 3 ft thick, there was a very pronounced maximum in the pollen analyses of pollen of ericoid type along with pine (Fig. 22). The ericoid shrubs, the heathers, lings and *Vaccinium* species are intolerant, as all gardeners are aware, of alkaline soils, but flourish on the growing surface of raised bogs where pines also characteristically grow. It was only shortly after this, during 1934/35, that we were able to recover detailed evidence of the

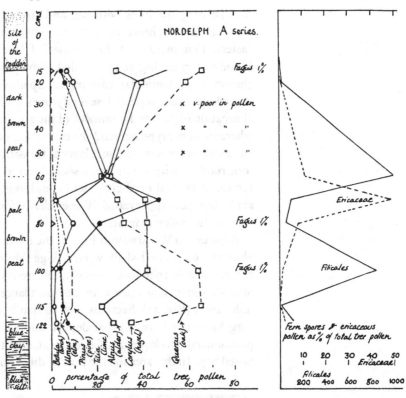

Fig. 22. Pollen diagram from the site of the Roman bridge at Nordelph (Fig. 27). The Upper Peat lying between the Fen Clay and the roddon silt provides evidence, in the high relative frequencies of pine, birch and especially ericoid pollen, that in its middle stages it had developed into acidic peat bog.

progress of this acidification at the marginal site of Wood Fen where it was quite clear that its maximum development took place after the maximal extension of the Fen Clay, although it was evident to a minor degree somewhat earlier. It was here also strongly associated with the local development of pine.

Elsewhere in the south-eastern Fens, even by the 1930s, peat-cutting, agriculture and drainage had so wasted the Upper Peats that very little evidence remained of any progressive acidification at this time. At a few places such as Wilton Bridge, Wicken Fen and Reach Fen the continuous deep peat extended upwards to a sufficiently recent date to embrace the period of the Upper Peat formation. This, corresponding substantially to the Sub-boreal climatic period, was identifiable in the regional tree pollen of our diagrams chiefly by three criteria, the progressive diminution of the frequency of lime pollen and the appearance of a consistent if low frequency of beech and, to a lesser extent, of hornbeam pollen. Using these criteria with evidence of level, it seemed apparent that here there had been no recognisable development to acidic peat bog. Very possibly this stemmed from the fact that this corner of the Fens remained subject to flooding by the several rivers here coming from the Chalk itself or Chalky Boulder-Clay thus keeping the Fens both wet and strongly alkaline. The situation was more favourable, however, in the Woodwalton Fen, Ugg Mere region of the eastern Fen margin. A long series of levelled sections and borings at Woodwalton during 1935 revealed above the marginal Fen Clay the very characteristic lenticular banding of *Sphagnum* peat→cotton-grass→ling, the whole cycle repeated several times and reflecting the characteristic alternation of pool and hummock of the growing raised bog. What was more, where this peat type overlaid the earlier fen peat dominated by reed, sedges and alder, there was a distinct layer of transition peat with a thick mattress of fern rootlets with birch twigs soon covered by those species of *Sphagna* typical of neutral or slightly acid conditions, before the onset of the very acidic bog peat supervened. The evidence for this phase was likewise to be seen in the pollen diagrams.

Adjacent to Woodwalton Fen on the north east there remained as late as 1822 the undrained shallow lake, Ugg Mere, the bed of which was still recognisable in 1936 by a deposit of calcareous marl with the abundant shells of fresh-water snails, just as in the much larger Whittlesey Mere less than 2 miles away (Fig. 23). Sections shewed that Fen Clay underlaid the whole of Ugg Mere but between that deposit and the shell-marl there was again a pronounced development of acidic peat, albeit of an aquatic type rather than raised bog. In this area too the peat acidification was clearly accompanied by

Plate *18*. *Eriophorum* (cotton-grass) peat from the Somerset Levels. The great fibrous tussocks of *E. vaginatum* are very easily recognisable and indicate acidic bog: they are the 'mabs' that make peat-cutting very difficult.

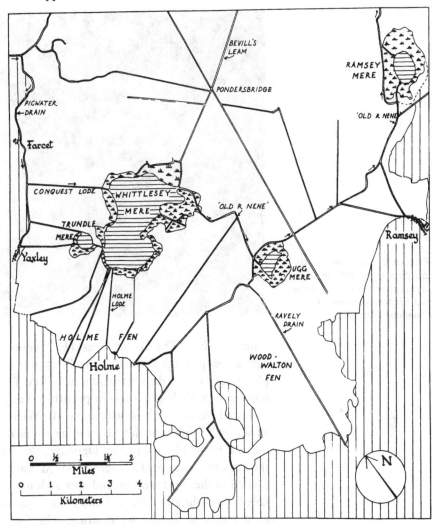

Fig. 23. The western Fenland margin as shown in Lenny's map of 1833. It gives the outlines of the still-undrained meres of Whittlesey, Trundle, Ugg and Ramsey with their borders of marsh or bog. The chief artificial waterways are shown and what was then known as the 'Old Nene'. This is the region of most conspicuous remnants of Fenland raised bogs; they were probably still growing hereabouts in 1833.

encouragement for the growth of pine, substantial stools of which, together with some yew and possibly juniper, occur freely rooted in the acid peat.

These proofs of the former presence of acidic bogs in the Woodwalton area enabled us to redress an unwarranted libel on the botanical knowledge, or even everyday country common-sense of an old Fenland observer Mr S. Wells, who had written of the Whittlesey area in his *History of the Bedford Level* (1830) that 'the turf moors are covered with such plants as the Heath, Ling and Fern, the *Myrica gale*, plants and natural productions, and a grass with a beautiful white tuft, called the Cotton Grass, are found in abundance'. Somewhat superciliously the naturalist Mr W. Marshall of Ely, writing in *Fenland, Past and Present* (1878), commented regarding this

Plate *19*. Cotton-grass, mostly *Eriophorum angustifolium*, fruiting freely as it grows in the wet *Sphagnum* moss of an abandoned peat-cutting.

statement that: 'The Cotton Grass (*Eriophorum angustifolium*) was 40 years ago, a conspicuous feature of the Fens, but it lingers now in only a few places. The rest of the statement is incorrect. The old surface of the Fen was no-where in Mr Wells' time, or probably at any time, covered with Heath, Ling and Fern, and those who knew Mr. Wells as intimately as the present writer, will think it no detriment to his acknowledged ability in all matters relating to the History of Fen Drainage, that he should not have regarded the Fens with the eye of a botanist.' Mr Marshall's experience was too much limited to the southern Fenland area with its maintained alkalinity, and he failed to consider that early drainage and peat-cutting could have removed most traces of former acidification. The fact that Miller conceded that *Eriophorum angustifolium* was formerly abundant and had much diminished since the 1830s, might well reflect a general cessation of peat digging by that time, since this cotton-grass is highly typical as growing abundantly in the trenches of old peat cuttings in acid mires throughout the British Isles (Plate *19*). Who knows what amounts of acid peat might not already by the early nineteenth century have been removed from the southern Fenland?

Our conclusions from the peat of the Woodwalton area were strengthened by the review of records and herbarium specimens of several typical raised bog plants collected in the Holme Fen, Whittlesey Mere area from 1834 onwards and by the persistence at Holme Fen of the cross-leaved heath, ling, cotton-grass and sweet-gale. When indeed after 1959 it was possible to extend a programme of boring and peat-stratigraphy to this area we were able to prove conclusively that large and entirely typical raised bogs had developed there and had occupied the marginal 2 or 3 miles of this corner of the Fenland continuously from as early as 2500 B.C. and were only marginally affected by the Fen Clay incursion. They have yielded not only pollen diagrams, reflecting the origin and continuity of the raised bogs, but peat full of, indeed composed of, all the characteristic mosses and flowering plants of this kind of mire. We shall return to the further evidence that the long pollen diagrams from Holme Fen and Trundle Mere also provide a prehistoric husbandry in the Neolithic, Middle Bronze and pre-Roman Iron Ages.

When Cyril Fox published his *Archaeology of the Cambridge Region* in 1923 he emphasised the extraordinary abundance of Bronze Age remains around the south-eastern edge of the Fenland, settlements, burials and hoards were frequent and stray finds still more common, extending, as Fox made clear, right into the peat-fens. It was apparent from comparison with his maps of other periods that in none of them was there anything like so intensive an occupation of the peat land. No archaeologists challenge these conclusions, so decisive are the differences. This argues a degree of dryness entirely consistent with the widespread acidification of the peat and with the extension of the Upper Peat far seawards upon the surface of the Fen Clay. Although the Sub-boreal climate may indeed have been to some degree warm and dry, the major effect was almost certainly due to the general regression in sea-level.

From the beginning the Fenland Research Committee were of course concerned to tie the major archaeological cultures into the sequence of Fenland history and, as we have noted, at a very early stage we were able at Shippea Hill to relate the base of the Upper Peat to an early Bronze Age occupation both at Plantation Farm and Peacock's Farm, that is to say on the sandy banks at each side of the Little Ouse. We sought of course to take advantage of the frequency of Bronze Age finds in that part of the Fens to make further correlation of the same kind and the skeleton of 'Nancy' at Southery gave confirmation of the Shippea Hill result. Other stages of the Bronze Age proved less decisive although there was nothing to contradict our early conclusion. Many of the finds came from beyond the limits of the Fen Clay penetration. This was so for the late Middle Bronze Age spear from Queen's Ground, Methwold, whose site of earlier recovery was investigated. It was possible to identify the original horizon of the find and, in 1934, it was possible to see from the pollen diagram that the horizon in the peat in which the spear was found corresponded with the Upper Peat further seaward, a conclusion supported by the growth at this horizon also of a good-sized yew tree, so that conditions were clearly not unduly wet.

Again in 1935 we investigated a Late Bronze Age occupation site on the sandy sloping Fen margin at Mildenhall Fen where the thin peaty cover and the waterlogged soil permitted a pollen-analytic sequence to be obtained. The results of this suggested an age younger than the Early Bronze Age and high values of pine pollen suggested the local growth of that tree during a dry phase of peat development such as was apparent at Wood Fen and was later to appear at Woodwalton Fen. It has indeed been a very general feature that the many Late and Middle Bronze Age finds adventitiously reported from the Fens, are so frequently associated with trees and quite often with pine. An interesting instance was the discovery in 1942, at Castle Hill Farm, Woodwalton, of a Middle Bronze Age palstave (Fig. 19*b*) which was found with the corner of the blade pecked into the trunk of a prostrate oak and apparently lost or abandoned in that position 6.5 ft (2.1 m) above the root-crown. Whilst there was no precise clue to the stratigraphy, the locality was still covered with shallow wood peat and the pollen analyses though not conclusive suggested once more a Sub-boreal age with local growth of oak, pine and alder as elsewhere in Woodwalton Fen.

A singularly exciting discovery was that by a ploughman in 1939 from the narrow belt of peat land separating the islands of Ely and Stuntney (Fig. 24). It consisted in a substantial bronze-founder's hoard with abundant worn specimens of bronze ribbed palstave, socketed axe, hollow headed spear and socketed gouge together with a broken sword fragment or two and several lumps of bronze metal. Of particular interest archaeologically were types of ribbed palstave, ribbed socketed axe, and faceted socketed axe that, with the

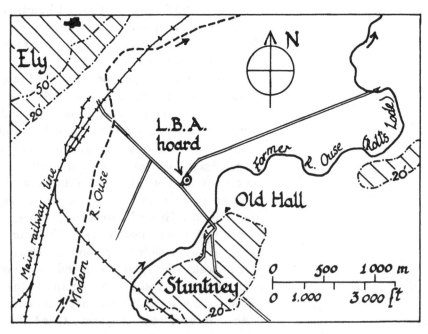

Fig. 24. Sketch map of the site of the bronze-founder's hoard from the peat off the flank of Stuntney island discovered during 1939. Note the natural (extinct) and artificial course of the River Ouse.

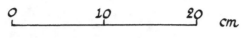

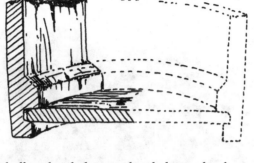

Fig. 25. Base of the wooden tub found at Stuntney, containing the hoard of damaged bronze tools and founder's ingots. The sketch shows (left) pieces of the recovered wall and base-board, and (broken line) an indication of the wooden vessel inferred from them.

hollow-headed spear-head, have also been found together in other Late Bronze Age sites (Fig. 19c). All these were inside the remains of a wooden tub that investigation shewed to consist of a body hollowed out concentrically from a single trunk of alder, and a base made of a flat board of alder wood also with bevelled edges set into a groove cut inside the body near its lower edge (Fig. 25). This site had the great advantage of shewing the base of the wooden tub *in situ* about 5.5 ft (1.8 m) above the surface of the Fen Clay. It was of course probable that the hoard had been inserted into the peat from a somewhat higher surface but the pollen analyses from the level of the tub, and from the peat still in the sockets of seven of the axes agreed in indicating that the hoard was to be referred to the closing stages of the Sub-boreal period, that is to say shortly before the onset of the second phase of renewed general flooding in the peat fens and deposition of semi-marine silts further seaward. It is satisfactory that we find here that the Late Bronze Age is stratified substantially above the horizon of the Early Bronze Age and falls later in the pollen-analytic zone.

Although other Bronze Age hoards have been discovered in the south-eastern peat fens some of them, such as those at Grunty Fen, of great importance and richness, were found either long ago or in conditions which did not allow a direct stratigraphic correlation with the broad Fenland story which we were reconstructing.

This was also the case with the prehistoric trackway that had been found some years earlier apparently part of a crossing between Ely and Stuntney. Although so close to a bronze hoard site as to suggest a possible link between the two, we were unable to recover any trace of the structure. We were happily much better placed at the Barway (or Little Thetford) site that came to our attention later in the same year as Stuntney. There had long been a strongly-held hypothesis that the line of the lost causeway was that by which William the Conqueror had secured his victorious entry into the Isle of Ely. Despite search no evidence of the hypothetical causeway or of conflict

Plate *20*. Line of the Late Bronze Age trackway that crossed the river Cam at Barway, the point at which the Isle of Ely most closely approaches the upland margin. The farmer has dragged up the main oak piles of the track to project where they were found. The age of the track is indicated by pollen analysis of peat where it meets the present river bank, by datable pottery sherds and a radiocarbon age of one of the piles of 610 ± 110 b.c.

anywhere along it had been secured, and some members of the committee, approaching the problem in a Holmesian way had found from the Ordnance Survey maps that in fact the closest approach of dry upland to the Isle of Ely was to be found at its south-eastern corner between Little Thetford to the west of the River Ouse and Barway to the east on a rise projecting north west from the Soham peninsula. The farmer of the ground east of the now heavily embanked river, after some careful approach questioning, made it apparent that such quantities of wood were present in a line across his field that a causeway had indeed been present in the conjectured position. Although the peat wastage and agriculture prevented immediate recovery of the upper layers of the track, the farmer at a later stage hauled to the surface a double line of massive pointed piles that extended down to the valley floor and had doubtless been the main supports of the causeway (see Plate 20). In order, however, to secure evidence of the trackway in relation to the peat stratigraphy where pollen analyses might be possible we made a small excavation where the track had disappeared under the edge of the river bank and where we could best count on its preservation. About 10 to 15 in (25–38 cm) of wood-peat appeared to have been surfaced by sand and this was followed by 'made-ground' of peaty clay with fresh-water shells. The pollen analyses indicated a Sub-boreal age and this was consonant with the presence of a few Late Bronze Age artefacts found in the peat close by, most probably in relation to the track. When radiocarbon dating became possible, an oak pile from the trackway was assayed, its outermost rings (the date of felling) yielding an age of 510 ± 110 b.c. This is at the very end of the Bronze

Age at a time when the worsening climate and possibly the approaching marine transgression would have been providing stimulus to the building of such trackways. The raised bogs of the Somerset Levels have provided many examples of Late Bronze Age wooden trackways similarly in response to flooding of the bog surfaces at much the same time (Plate *21*). It is a pity that peat destruction has gone so far in the Fenland that the chance is remote of recovering other examples and learning more of their purpose and manner of construction. This being so we are particularly grateful for the published record of a track that formerly crossed Woodwalton Fen, even though we are ignorant of its age.

It was a great pity that members of the Fenland Research Committee never chanced to make contact, in his later years, with William Henry Edwards of Lotting Fen, Hunts., for his acute and accurate powers of observation and his life-time of work in the Fenland would have provided them with invaluable leads in their investigations. I noticed this especially when I found, in his daughter's delightful book, *Fenland Chronicle,* his account of the discovery by 'old' Jackson of a corduroy trackway in the peat, apparently running from Castle Hills Farm to Honey Hill, Ramsey Heights across Wright's Fen, now the southern half of the nature reserve of Woodwalton Fen. 'Will 'En' gives clear description of the construction of a track made entirely of cut pine packed very close together and pegged down

Plate *21*. Opposite the figure 11 on the staff one may see the pale timbers of a light wooden trackway of the Late Bronze Age, exposed in a peat cutting at Westhay, Somerset. It is easy in such circumstances not only to excavate and determine the construction of the track, but to secure for pollen analysis and radiocarbon dating all requisite samples free from contamination: a vertical series has been taken here. The peat here is highly-humified *Sphagnum*–ling–cotton-grass peat.

with stakes 6 or 7 ft long and pointed at the bottom by 'some sort of edged tool'.

In the light of our subsequent studies in the region it is perhaps permissible to conjecture as to the age of what I like to think of as 'Will 'En's track'. By 'edged tool' was almost certainly implied a metal one, and the work was therefore of Bronze Age or later. Its considerable depth from the peat surface of that time makes an age as great as the Middle or Late Bronze Age not unreasonable, and there have been recovered from the peat in or close to Woodwalton Fen, no fewer than four separate bronze axes, three of the Middle Bronze Age and one of the Late. Moreover, it is now apparent from pollen analyses, checked by radiocarbon dating at Holme Fen close by, that about 1450 b.c. there began a phase of woodland clearance for agriculture on the neighbouring uplands so that certainly in the Middle Bronze Age there was a local population engaged in tree-felling, possibly a numerous one. It is distinctly unusual in this country to have a prehistoric wooden causeway built of pine, and this suggests considerable local prevalence of the tree, which is much tougher to cut than alder or birch and which is too heavy to invite further carriage than is needed. However, the studies made in the 1930s at Woodwalton and Ugg Mere revealed great quantities of pine that grew not merely in the basal buried forest where its frequency is relatively small, but on the peat surface itself, both before and after the Fen Clay transgression was at its height. The suggestion is strong that at some time when flooding made passage across the Fens more difficult, it was expedient to facilitate crossing from one low upland to the next by constructing trackways of the most readily available local timber, to wit, the pines growing on the wet bog surfaces. What satisfaction there would be to recover for carbon-dating some fragment of this track perhaps from beneath a drove or dyke bank still untouched by peat-cutting or wastage.

9

Iron Age hiatus, roddons and Romans

Already in 1920 the comprehensive distribution maps newly constructed by Fox for the region of Cambridge had unmistakeably brought out 'That the southern fens and their eastern borders from Quy to Fordham, so rich in remains of the Bronze Age, are in the Iron Age almost entirely barren', and this virtual abandonment of formerly populous Fenland has later been demonstrated to be equally typical of the whole marginal Fenland. So empty of Iron Age finds is the Fen region save for a very occasional single lost coin or sherd, that comment on the period is idle save for speculation on the possible reason for so great a withdrawal. Far the most likely cause is a widespread return of waterlogging after the dryness that in Sub-boreal time had made the peat subject to colonisation alike by trees and humans. The reason for the waterlogging, in part at least, may have been the shift of climate towards a cooler, wetter and generally oceanic type, such as took over in all north-western Europe after a transition about 800 to 500 B.C. On the other hand it remained to be ascertained whether the resumption of severe marine transgression evident in Romano-British time might not well have begun some centuries before the actual Roman occupation.

By the 1930s there were beginning to appear in many parts of the Fens, meandering banks of pale buff-coloured silt, rising so gently above the general ground level that it took an experienced eye to discern them, especially as one so seldom could find a view point more than a few feet above the soil surface. These banks were the structures recognised by Gordon Fowler as extinct rivers and already named by the fen inhabitants 'roddons' or 'rodhams', a term applied to equivalent structures in the Netherlands. They were becoming apparent through the wastage of the peat (formed after their deposition) that had covered them over and was now disappearing thanks to the improved drainage which accelerated in the more intensive cultivation of the war effort. The stability of the silt banks had long been appreciated by the fen people however, and farmsteads in a roddon landscape almost invariably occur on them and not on the peat (Plates 22 & 23). As the peat wastes, the straight Fen roads built upon it gradually sink and, if they cross and recross a large roddon, develop a switchback that soon

Plate *22*. The Little Ouse roddon between Shippea Hill and Littleport, showing the considerable width of the silt ridge, its pronounced convexity and the manner in which farmsteads commonly have been sited along its crest (photographed early 1930s).

Plate *23*. Roddon of the natural River Little Ouse photographed from the air in 1964, looking upstream between Littleport and Shippea Hill. The pale silts of the roddon contrast strongly with the black peat from which they now increasingly project. The dark central line along the crest of the roddon is the peaty trace of the fresh-water stream that marked the last stages of the natural waterway. Peat wastage has now so advanced that it reveals the pattern of salt-marsh creeks upon the surface of the Fen Clay, some 2000 years older than the roddon system. (Photograph by Department of Aerial Photography, University of Cambridge.)

draws a car-driver's attention to the structures he is crossing. The crops too respond to the different soil properties of nutrients and moisture retention by shewing the roddon course by changed height, vigour and colour. This feature is also true of a small central depression running along the top of the roddon crest, the last vestige of the open waterway, sometimes with a local peat formed in its own channel and sometimes with a locally persisting name shewing it to have been a recognisable waterway in recent time, as is strikingly the case with the 'Old Croft River' marked as an open stream on Ordnance Survey maps through Welney village and northwards to Upwell.

These roddons were only a part, though the most exciting part, of the extinct system of Fenland waterways whose recovery and documentation was Fowler's very special contribution to Fenland research. It has to be realised that his work of mapping and transfer to the six-inch Ordnance Survey map was done under fearfully daunting field conditions, walking every yard of the ancient rivers on his own two feet (one of them *ersatz* at that). Air photography which came extensively available only later served mainly to confirm, though certainly also to extend, conclusions Fowler had already reached, especially as demonstrating that the roddons formed an extensive and intricately branching system of natural meandering streams converging from the southern Fens upon a great natural outlet via Welney, Outwell and Wisbech to the sea. Of this system little now is occupied by an active river: it has all been short-circuited and substituted by a maze of man-made drainage cuts of all sizes and periods.

Fig. 26. Map of the eastern end of the Roman causeway from Denver to Peterborough, crossing the ancient natural estuary of the Fenland, the seventeenth-century Bedford Levels and the junction at Denver of the *Cut-Off* and *Relief* channels with the Bedford rivers and the River Ouse. Four sites of research excavations refered to in the text are shown ringed.

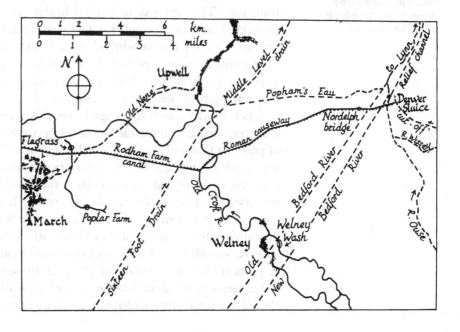

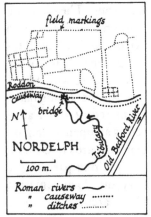

Fig. 27. Sketch map of part of the Roman causeway to Peterborough near to its eastern end at Denver. It shows the pattern disclosed by air photographs and excavations made by E. J. A. Kenny. The pollen diagram of Fig. 22 was made from samples taken below the eastern abutment of the bridge by which the Roman road crossed the tributary stream that is now a roddon confluent with that carrying the road itself. The field pattern is clearly disposed in relation to the roddon bank system.

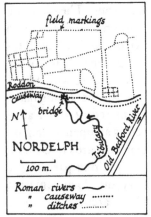

(map labels: field markings; Roddon; causeway; bridge; N; NORDELPH; 100 m.; Tributary; Old Bedford River; Roman rivers —; " causeway ....; " ditches ....)

The maps Fowler published in 1933 and 1934 had a startling impact and played no small part in galvanising the Fenland Research Committee. The borings put down during the Plantation Farm investigation crossed the roddon deposits of the natural Little Ouse and yielded information on its shape and altitude in relation to present sea-level. This was especially important since the foraminiferal analyses had made it clear that the roddon silts were deposited in almost marine conditions, quite certainly from a tidal channel that had transported them upstream from the sea. Furthermore, the indications were that the upper silt at least was of Romano-British age. Fowler was already aware that beyond Welney the silts of the main estuarine roddon merged into the area of continuous upper silts stretching thence right to the coast. In 1933 I visited with him a site at Three Holes, two miles north north east of Welney where there is a Romano-British settlement whose abundant remains are scattered in the upper silt, that there typically overlies the characteristic Upper Peat, itself above the Fen Clays.

It had long been known that a gravel causeway built by the Romans crosses the Fens from Denver to Whittlesey, continuing no doubt to Durobrivae and Caistor just up the Nene from Peterborough. It was apparent that it sat upon the south flank of a large roddon and air photographs shewed that on the same side the roddon had been joined by a tributary stream (Fig. 27). If then the roddons were flowing channels in Roman time, how did the causeway builders respond to this interruption? To resolve this query was the prime purpose of the excavations made by Dr Kenny who was able to shew the Fenland Research Committee *two* Roman responses. The first was to build a bridge whose abutments were now uncovered, and the second was to construct a ford alongside, from the bottom of which Romano-British objects were recovered. It would have been difficult to have proved more convincingly that the roddons were indeed active in Roman times: the other valid conclusion from this work took unreasonably long for us to appreciate!

When Fowler was first engaged upon his mapping programme he had been very struck by the evidence for the very great contraction of thickness of peat that followed directly upon the drainage of Whittlesey Mere, and he was quick to associate this with historic evidence he found for repeated lowering of drains and their outfalls in the South Level as the level of the peat itself went down. On this phenomenon he based his explanation of the present-day convexity of the roddons. He took their silts to have been deposited in the concave bed of a channel cut into the Upper Peat, assuming them reasonably to have been thick in the middle and thinning on the banks. A proportionate contraction of the peat subsequently would lower the lateral deposits far more than the centre below which the peat was already thin. Thus the silts would have assumed their present shapes.

What stuck in the gullet was the thought of the Romans building a gravel causeway in the bottom of Kenny's roddon stream at Nordelph. In fact the hypothesis was wrong in attributing all lowering of peat levels to contraction; although compaction, following directly on a sudden drainage is dramatically sudden, it is quite unimportant as a continuing process. So long as peat is waterlogged it remains more or less intact, but as water-level is lowered by drainage, the access of oxygen to the top layers exposes them to bacterial and fungal attack and to direct chemical oxidation, so that the peat *wastes* by immediate or indirect conversion to carbon dioxide for the main part. Thus the lowering is not proportionate through the whole peat thickness as in contraction but represents a progressive paring away from above. This is clearly shewn in the pollen diagrams that get truncated from the top and even by the exposure first of the late archaeological material and later that of earlier cultures. The possibility was thus raised that in fact peat wastage was doing no more than reveal the original shape of the roddon, a suggestion given colour by the fact that on the north-west German coast the Jade and Weser estuaries at the present day have just such raised tidal banks standing up above the marsh and fen that lies behind them. These banks narrow upstream as the tidal range becomes less and the fens behind are so shielded that they carry big areas of raised bog. There was evidently the clear need to excavate and measure a section through a roddon so as to observe its relation to the underlying peat and to see whether it bore any signs in its stratification of the collapse of its silt layers as required by the original hypothesis. An opportunity for this exercise was presented in 1933 by the cleaning of a deep ditch section across a medium-sized roddon at Poplar Farm, about 1 mile (1.6 km) south-east of March (Fig. 28). The wind-scoured faces of the silts in the roddon were clearly laminated in coarser and finer components and their uniform horizontal bedding shewed no trace of subsidence or fracture. The very wide silt flanges moreover covered a uniform thickness of the Upper Peat no thicker towards the edge than to the centre of the roddon. These flanges still rested on the flat upper surface of the Fen Clay just as they had probably been at the time of deposition. We could not escape the conclusion that the roddon silts were deposited as they had been found: that is to say they were the natural raised banks produced by flooding of a tidal river, a conclusion that Fowler readily accepted.

We now had a solid basis from which to reconstruct at once the relation of these waterways to river-side occupation, road construction and water-transport in Roman time. One could now readily perceive how at the conclusion or near the conclusion of their building up by tidal action, they would admirably serve as the foundation for such causeways across the Fens as that we had already excavated at Nordelph, and it was possible, as it was

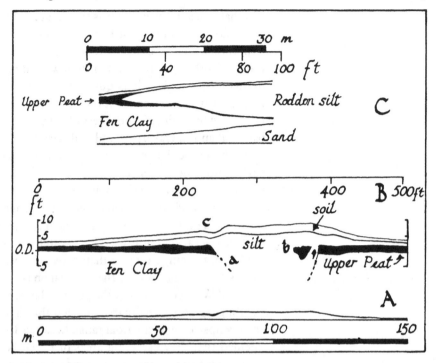

Fig. 28. Measured section through two roddons; A, that at Shippea Hill given in natural scale, and that at Poplar Farm, near March shown in B with fourfold vertical exaggeration of height and in C as a natural scale profile. In B, (*a*) indicates the channel eroded through the Upper Peat and Fen Clay by the active roddon stream, and (*b*) a large erratic of peat that was cut free and left in the roddon silts: (*c*) shows the final position of the reduced open waterway on the crest of the roddon.

later to be proved, that they might actually carry settlements on their raised banks. In fact evidence pointing to this latter conclusion was close at hand. Not only were there clay 'squeezes' of a kind common in Romano-British wattle-and-daub present in the Poplar Farm roddon silts, but that roddon was shewn to be a tributary into an enormous roddon that less than a mile away runs from the east towards March and carries the significantly named 'Rodham Farm'. The slopes of this great structure are thickly strewn with ditches and hearth sites attributable to the same period, and what is more it carries on its flank the Romano-British gravel causeway that began at Denver on its way towards March and Whittlesey.

So far so good, but we had now disclosed another unsuspected field of enquiry. The great Rodham Farm roddon could easily be observed over at least 4 miles (6.4 km) to follow so straight a course that it must necessarily have been artificial, and thus presumably was a channel cut from the confluence with the main Fenland estuary before or during a halt in the marine incursions, which nevertheless at some stage, by accident or design of man, invaded it so that it became tidal and built up its stratified raised banks just as the fully natural roddons have done. Subsequently it has been revealed that a roddon bearing a Roman gravel causeway apparently terminates on the eastern margin of the gravel island of March, at Flagrass where there are clear signs of extensive Roman occupation. A radiocarbon dating sample of the peat just below the laminated roddon silts gave a date shewing that the roddon could not be older than 1115 ± 110 b.c. Whether and when the roddon channels were navigable remains an open question, but one that it is important to have in mind in considering the vast extent of other known Roman waterways in the Fenland.

It will now be apparent what should have been deduced from the Nordelph data for the eastern section of the Roman causeway from Denver. Here too the roddon that carries the track is straight to a degree that no natural river is (see Fig. 26): it too must have been artificial, and if so, the deduction is of the digging of a Roman canal transversely across the Fens from Denver via March to Peterborough, a transport route of very great convenience. The trackway, useful as it was and made up into a durable road, was not the primary consideration but followed as a matter of taking advantage of the dryness of the elevated roddon banks.

The clinching evidence that the roddon levées were actually lived upon during the period of their formation was elegantly provided in field seasons of 1936 and 1937 by the excavations directed by C. W. Phillips where the main estuary of the Fenland crossed the washes between the two Bedford Levels close to Welney. Sections were made in an oxbow of the main channel that was almost cut off in Roman times but still subject to tidal penetration so that it was itself a roddon 850 ft (260 m) wide and 5 ft (1·5 m) high. Phillips has vividly conveyed the appearance of a typical Roman sherd of Caistor ware found at the site, a product of the extensive potteries in the Nene valley in the neighbourhood of Castor, Water Newton and Wansford. 'Here we have a more ambitious subject, the destruction of the Lernean Hydra by Hercules. The figure of Hercules brandishing his club is shewn with considerable vigour and success, but the artist's limited experience failed him when it came to the Hydra, so that the efforts of the hero to dash to pieces a kind of hybrid between an Aunt Sally and an octopus seem disproportionately great.' So much for provincial classicism! The painstaking excavations shewed that there were two periods of occupation, the earlier from the late first to late second centuries A.D. and the later from about A.D. 250 into the fourth century. The two occupations included dwellings actually built upon tidal roddon silt and a band of this material 6 ft (1·9 m) thick in the channel bed separated the two occupations. There was moreover an indication of occupation through the inundation in that the settlers seemed to have taken steps to meet the onset of the submergence. Phillips recalled that at the Nordelph causeway site also, a silt deposition phase had intervened between the two succeeding gravelled surfaces of the road. The Welney Wash roddon was quite typical in the ecological implications of its foraminiferal assemblages and in the presence on its crest of the small channel active in its last fresh-water stages long after Roman time. The broad morphology of the roddon system was dramatically displayed to aerial view in the extensive floods of 1947 when only the broad winding silt bank still stood above water, its latest narrow channel evident, as also the parallel ditches bounding a Romano-British droveway, and ditches round the small square fields of the contemporary agriculture.

A good deal later the site was reopened in 1960 by Dr D. M. Churchill in an attempt to secure a reliable radiocarbon age for the first onset of marine conditions. This is far less easy than it appears and there has been great variability in the dates obtained directly underneath the silts especially at more seaward sites. One has to reckon on the one hand with possible erosion of the top peat surface by the incoming tides, and on the other with growth downwards of stems and roots growing on the later marsh surfaces. Here at Welney samples were taken where there was no visible discontinuity between Upper Peat and lower silt, and no present trace of downward penetration of reeds. All the same, allowance was made for a possible downward penetration of soluble humic material. When every allowance had been made, at most 300 years, it seemed very probable that here the tidal inundations commenced between about 600 and 300 b.c. This is a result of particular importance since it shews that the phase of extreme wetness that forbade all Iron Age settlement in the Fenland, save for that on a few gravel islands, had its origin in the beginnings of the marine transgression that was to culminate between the third and fourth centuries A.D., and after which the Fenland remained utterly inhospitable right through Anglo-Saxon time.

As Fowler continued his investigations into the extinct waterways of the Fenland his conviction grew that the Romans had been responsible for a quite unsuspected length of drainage and navigational cuts often of considerable length and cross-section. The first such wholly within the Fenlands was the portion of the present River Great Ouse, or Lynn Ouse that runs from Prickwillow northwards to Brandon Creek, the confluence with the artificial River Little Ouse and thence, more conjecturally, to Stowbridge. Its obvious straightness and freedom from the meanders of natural Fen rivers at once proved its artificiality. Moreover, at its southern end it lies quite transversely to the natural channel of the Wisbech Ouse, the great roddon carrying the waters of the natural Little Ouse coming down from Shippea Hill, joined with those of the rivers Granta and Lark coming from the south. Since the Little Ouse roddon was already established as Roman, the artificial cut had to be itself Roman or later. To date it more closely Fowler secured the co-operation of the drainage authorities then engaged in cleaning out the channel: pottery sherds recovered from the dredgings were kept separately mile by mile along its length and every section proved to contain such abundance of Roman ware that it could only have come from Roman boat traffic upon the river. It could not have been derived from the ground surfaces traversed by the cut for most of the route lies through deep fen peat quite unsuitable for settlements and lacking all trace of such. As Astbury has pointed out, there is a lot of evidence pointing to transport of Roman pottery along the Fenland waterways, as for instance in the abundant finds from the River Granta at Clayhythe and Upware and from the roddon that lies under the upland at Stuntney, where indeed the abundance of Castor ware suggests possibly a regular site for unloading.

Once a Roman origin has been proved for so large a work as the Prickwillow–Brandon Creek (Stowbridge) cut, the way has been opened to speculation whether many smaller channels, such for example as the artificial River Little Ouse from Shippea Hill to Brandon Creek, and even the main Fen lodes such as Burwell, Upware and Reach Lodes might not also be of this age. Exciting as these suggestions may be, proof or even very strong circumstantial evidence is mostly lacking, though we note as an exception the old slade on Reach Fen (see Chapter 10). What we also remain tiresomely ignorant about is the extent to which the Romans made navigational use of the Fenland river system when it was still tidal and flowing between raised silt banks. By the same token we do not know what, if any, was the link between the Prickwillow–Stowbridge cut and the tidal channels that traversed the extremely populous silt lands right round the Wash.

We are in an area of greater certainty with the great navigational canal, the Car Dyke, dug by the Romans all of 70 miles from the Cam at Waterbeach northwards to Peterborough and finally to the Witham at Lincoln having crossed *en route* the east-flowing rivers Great Ouse, Nene, Welland and Glen. Stretches of the waterway are still easily recognisable; alongside the boundary of Waterbeach aerodrome beside the main Cambridge to Ely road: where that road turns north north east along Akeman Street, the not-too-impatient motorist can see the straight line of the Roman canal keeping its north-east direction through the cultivated fields towards the Old West River and Earith, and those fields in fact still yield abundant Roman artefacts. From Earith, where the Great Ouse enters the Fenland, there is no trace of an artificial cut and it is assumed that transport thence followed the natural course of the Ouse, now a large roddon, north as far as Benwick and thence by some river or canal route not certainly identified, to Peterborough where a clear cut channel of size and shape comparable to that of the southern section leads off northwards across the Fen margin where it has been recognised over large distances. It was suggested by the great pioneer archaeologist Stukely that the purpose of the Car Dyke was to supply the legions in northern Britain with corn from East Anglia and this conjecture has been generally accepted, the alternative view that it had been made to safeguard the Fenland from discharge of upland flood-water

proving untenable on any close consideration. Excavations of a well-preserved portion of the dyke at Cottenham shewed an early period from the
first century A.D. when it was kept open as a canal, and then a stage, probably
in the fourth century when the settlers on either bank, finding it a nuisance,
had blocked it by a transverse filling.

When we have reached the Romano-British period we are close to the
limits of usefulness of the stratigraphic–geological approach to archaeology
that was the hallmark of the Fenland Research Committee, for deposits of
this age lie at the very top of the layer-cake that constitutes the infilling of the
Fenland basin: we can then no longer call on the advantages of setting
human cultures in a sequence of geological strata, nor can we make use of all
the dependent lines of evidence of past fauna, flora and climatic shift. Pollen
analysis is, in any event, of greatly diminished value in these uppermost
strata except as index to the progressive modification of the natural forest
cover by human interference.

These various limitations made themselves strongly apparent to the
Fenland Research Committee and their successors when they came to
consider the great belt of silt land, some 15 miles in breadth, that surrounds
the Wash and was deposited in considerable thickness above the Upper Peat
as we saw in the St German's excavation. The silt surface, today reaching
levels of 10 to 15 ft (3 to 4.5 m) above Ordnance Datum, was readily
observed from the earliest days of air photography, to be thickly strewn with
traces of occupation, soon resolved as of Romano-British age (Plate *24*),
associated with branching patterns of creeks that suggested settlement
during a phase of extensive development of upper salt-marsh.

Detailed investigation of the region was pursued over many years, most
notably by Dr S. J. Hallam, the results finally appearing in a special Royal
Geographical Society Memoir in 1970. Research now made use of greatly
improved standards and full cover by aerial photography for which
conditions were almost ideally suitable whether the ground is recently
ploughed or under crops: the ancient field marks are as clear as the modern
features thanks to contrast in colour and water-holding powers between the
pale silts and the dark peaty depressions. The 'magnificent' detail was
transferred to large-scale maps and there followed stages of field survey,
excavation of key sites and stratigraphic enquiry where possible. A very
early bonus was the discovery that features of the Middle Ages in drainage
and surveying were so rectilinear as to be fully recognisable, leaving for
interpretation the imprint of no more than the four centuries of the
Romano-British occupation, and here the abundance of datable sherds on
and just below the surface was such that it was possible even to trace the
vicissitudes of the region half-century by half-century. The dense pattern of
watercourses, dyked droves, field and farm boundaries, the siting, grouping

Plate *24*. Aerial view of Romano–British earthworks on the silt Fen, 1.5 miles (2.4 km) south south west of Holbeach St John's, Lincolnshire. The South Holland Main Drain bisects the picture: on the north side of it is 'Shell Bridge' settlement site, and to the south of it 'Somerset House' site. Excavations have confirmed widespread first century industrial activity, and abundant settlement debris from first to fourth century A.D. The markings include droves, field boundaries and occupation enclosures with small rectangles within a larger rectangle. The long straight parallel marks are of later ditches. The site was under grass in 1960 when the picture was taken. (Photograph by Department of Aerial Photography, University of Cambridge.)

and size of farmsteads were now all capable of analysis to yield information on population size and its sustaining economy. It soon emerged that the bulk of the 'silts' having been deposited in the pre-Roman Iron Age, their clay surface had been first occupied during the later part of the first century A.D., when settlement was 'casual and small-scale', but that settlement had thereafter become 'rapid and official', reaching its *floraison* in the second century. A phase of renewed deposition of tidal silt led to depletion of settlement during the third century and after a brief rejuvenation, the Marshland was deserted after A.D. 450 and right through Anglo-Saxon time. Analysis makes it clear how large a farming population the Roman silt lands sustained: the farms in total as numerous as today, at first tended to occur singly and later were aggregated into tiny hamlets. The dykes were evidently regularly maintained and were liable all along to invasion by tidal water. Although some cereals were grown, mostly spelt wheat and hulled barley, this was primarily for local subsistence. The main products appear to have been sheep and cattle, the former particularly on the upper salt-marsh that does not harbour liver-fluke, and the cattle on the somewhat lower and lusher ground extending inland to the peat Fen margins. Dr Salway writes 'we may envisage the later Romano-British fenland as cattle-ranching country, not unlike parts of the present-day Camargue, with large farm-buildings housing stock standing in their yards among drained and undrained pasture, and with arable only in the driest parts of the region'. It is conjectured that hides and wool were among the most important products supplied to the military north, rather than corn, as had been hitherto supposed.

Another factor of considerable commercial importance was the winning from sea-water of salt which was essential for the preservation of meat and important in the curing of hides. The recognisable crude briquetage of saltings with its surfaces burned orange-red, is abundant throughout the region, and it is strongly suggested that the necessary fuel was peat either brought from nearby peat fens, or recovered by digging down to the Upper Peat that underlies much of the silt. Salterns of similar age occur in quite comparable situations near the peat–clay junction on both the Lincolnshire coast and in the Somerset Levels, and there is written evidence that later on, in A.D. 1200, turbary rights were actually associated with the lease of salterns.

It is evident that the salt industry must, like the agricultural export trade, have made substantial demands on transport and both roads and canals have been identified in this survey of the Romano-British silt landscape. It is interesting that as with the Rodham Farm canal further inland, so here the resumption of some marine invasion followed cutting of straight waterways, overlaying the banks with further deposits of tidal silt.

There was of course no natural break in continuity between the main coastal belt of silt land and the landward extension along the main Wisbech–Welney estuary into the branching system of roddons penetrating the peat Fens. As we have seen, the pattern of Romano-British occupation in both territories followed a broadly conformable course. Throughout the black Fens at the present day the peat has almost everywhere wasted away to far below the Roman level, so that almost no chance remains of finding Romano-British objects stratified into undisturbed peat. This was particularly unfortunate with the very important Roman settlement investigated by P. Salway in 1961/62 at Hockwold-cum-Wilton, conjecturally identifiable as the 'Camboritum' of the Antonine journeys. The stratigraphic deficiency holds quite generally save for the important exception of the former meres that are described in the next chapter, peat that must have formed over the roddon flanks has disappeared since the operation of effective drainage and so has the peat that doubtless developed above the landward margin of the coastal silts.

IO

Extinct meres and shell-marl

On the beautiful Fenland maps of the last three centuries appears a feature lacking in those of the twentieth century and of today. This is the considerable number of fresh-water lakes or meres, for the most part sitting beside the main rivers in their upper courses across the peat fens. These ancient meres are now extinct, victims of the increased effectiveness of the Fenland drainage and only their names appear on the Ordnance Survey maps. The old maps shewed that they were of all sizes downwards from Whittlesey Mere, then ranked as the second largest of all English lakes. Their names strike a very familiar note in the Fenland story, Whittlesey, Trundle, Ugg, Brick, Ramsey and Benwick meres in the system of the ancient 'Nene', Streatham, Soham, Willingham, Harrimere in that of the Ouse or its tributaries and many more so small as to have been forgotten. The meres were an important source of wealth in the economy of the undrained Fens, and their fisheries were carefully assessed and allocated, as may be seen for instance in the Domesday records. After deliberate drainage, mostly in the middle of the nineteenth century, their beds were put under the plough. In the picturesque phrase of W. Wells about the drainage of Whittlesey Mere: 'The wind, which in the autumn of 1851 was curling the blue waters of the lake, in the autumn of 1853 was blowing in the same place over fields of yellow corn.' Despite the radical changes in the landscape the sites of the meres remain dramatically obvious by the strikingly white colour of the lake deposits in contrast with the surrounding black peat (Plate 25). The sediments are of fresh-water chalk deposited through the photo-reduction of dissolved carbon dioxide by green water-plants submerged in the shallow clear water. In particular one can recognise the stems, branches and 'fruits' of the stoneworts (*Chara* spp.) the largest of our fresh-water algae, and the crumbly porous mass is sown through by hosts of the delicate shells of fresh-water snails of a number of common species in the genera *Pisidium, Sphaerium, Bithinia, Succinea, Limnaea*, etc. The lakes are thus readily detectable from the ground, and are still more evident from air survey. This reveals not only one extinct mere even larger than Whittlesey, Red Mere by the River Little Ouse, but

Plate 25. Vertical air photograph shewing the greater part of the former Whittlesey Mere. It is strikingly indicated by the white marl of the former lake surrounded by acidic peat bog, now largely cut and drained, less so at Holme Fen, south of (*b*). On its northern and eastern margin the mere is limited by the winding silt banks of a natural river (*a*) possibly the Nene. At (*c*) there are relict creek systems draining east and representing the surface of the Fen Clay now revealed by wastage of the Upper Peat. (Photograph by Department of Aerial Photography, University of Cambridge.)

innumerable minor pools scattered in the peat Fens. It seems apparent from suitable sections in the least drained regions, that these minor pools eventually filled and by the process of vegetational succession became converted to peat fen.

It is thus evident that the formation of the meres represented virtually the latest geological phase in the history of the peat Fenland, but it clearly remains to fit it to the story that we have already carried as far as Roman time, and to consider what were the circumstances that caused so much open water suddenly to appear. Skertchly in 1877 had already commented on the remarkable lack of human artefacts in the shell-marl, and those that were subsequently found, such as the silver censer found in the drainage of Whittlesey Mere, were such as followed naturally from the use of the lake for medieval boat traffic, and did not serve to suggest a date of origin.

Although the Fenland Research Committee's excavations at Shippea Hill encountered only an Upper Peat layer too wasted to contain any record of the mere stage, we were at the same time engaged in work on the Fenland margin near Woodwalton Fen, where this disability did not apply nearly so strongly. The long deep section recently cut along Green Dyke in Lotting Fen (Fig. 20), transected the bed of the former Ugg Mere that was still shewn as open water in drainage maps as late as 1833. Here it was easy to establish the sequence of strata. The Lower Peat and above it the Fen Clay extended everywhere below the mere, as it was possible to shew it also did at the neighbouring smaller Brick Mere. Above this was a continuous layer of Upper Peat and then the shell-marl of the lake-bed with its typical fresh-water mollusca. Although continuous pollen analyses were made through the deposits they were of less use for establishing the ages of the upper layers than for shewing, from the local pollen, the ecological nature of the conditions of deposition. In this they strongly supplemented the positive evidence of the larger plant remains, stems, leaves and fruits, of the Upper peat-bed. These together proved the altogether dramatic overturn of ecological conditions represented by the creation of the mere.

It was easy to trace by the diatoms, foraminifera and pollen alike how the surface of the Fen Clay became in turn salt-marsh and shallow fresh-water, but now the area received so little calcareous drainage water that it was invaded by the calcifuge *Sphagna*, and became a wet *Sphagnum* bog whilst in more marginal situations, as we have seen, raised bog carrying birch and pine trees was able to develop. This régime of relative dryness and freedom from flooding was now abruptly terminated by conditions creating not only lakes, but highly calcareous lakes in the same area. It was not until some years later that we were able to get a radiocarbon date for a sample of the acid peat we then collected from below the shell-marl. The radiocarbon age of 1310 ± 110 b.c., however, merely confirmed the general sequence we had already deduced.

The same reversal from oligotrophic acidic bog to calcareous open water was demonstrated even more strikingly at the same time in our investigations at Trundle Mere, a neighbouring marginal site outside the range of the Fen Clay, where birch woods on strongly acid peat were directly submerged by the newly forming lake.

Dramatic as these results were, we nevertheless had to return to the South Level for demonstration of the essential nature of the meres and their origination. In 1947, at the suggestion of Gordon Fowler, Anthony Vine, as already reported, had mapped on the ground the full extent of the great area of shell-marl lying east of Shippea Hill, that is traversed by Cross Bank and bounded on its northern side by the abandoned natural channel of the River Little Ouse and by Lakenheath Lode. This very convincingly shewed that there must formerly have existed here a fresh-water lake, larger than Whittlesey Mere, although it did not appear on any known map and no drainage record exists of it. Fowler suggested that only the local name

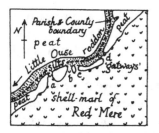

Fig. 29. Map after J. N. Jennings to show the close lateral contact between the north-western flank of Red Mere and the silts of the Little Ouse roddon. The presence of open access of water between the two was proved by borings at the four 'gatways' mapped *(a–d)*. Thus the mere was proved contemporaneous with the roddon, i.e., Roman (see also Fig. 15).

'Redmere' perpetuated its existence, and this is the name now generally adopted for it.

It was to Red Mere that J. N. Jennings turned after his studies on the Norfolk Broads had suggested to him a possible strong analogy with the lake system of the Fenland. It was already clear that all along the north-western margin of Red Mere the lake deposits fitted closely against the meandering bank of the Roman roddon of the Little Ouse, and it now was shewn that in four places the shell-marl extended through a lateral channel in the top of the roddon (Fig. 29). Having wasted less than the surrounding peat the shell-marl now stood some 3 ft higher and so remained as four ridges across the roddon bank and some 50 ft (16 m) in width. Close-set borings proved everywhere the sequence, Lower Peat, Fen Clay and Upper Peat already seen at Shippea Hill, but here the Upper Peat stretches continuously below the lake-marl. The roddon silts were shewn to taper out so completely below the lake-marl that undoubtedly the 'Roman' transgression had achieved its maximum lateral extent before the lake came into existence (see Fig. 28). It seems quite apparent that the mere was created by the ponding-back of fresh-water behind the raised silt-banks, and Jennings had already made a good case for the Norfolk Broads having been similarly caused initially by the constriction of the river in banks (ronds) of tidal clay. He was able to point to the way in which the meres at Ramsey and Benwick occupied a similar lateral relationship to the natural channel of the 'River Nene' represented by yet another great roddon. Since there is no trace of salinity in the shell-marl and since so much of the roddon flank extends below it, one has to suppose that the meres came into being only when the roddon banks had reached full height, and indeed the marine transgression was past its maximum. We know that the latest settlement of the roddon slopes took place both at Welney and Rodham Farm during the third and fourth centuries A.D., so that this may reasonably be taken as the date when the fresh-water lakes originated, even though more general waterlogging was prevalent during the transgression itself. It is odd and tantalising that we have so little information of when, between then and the seventeenth century, so large a lake ceased to exist. Certainly to have formed upwards of three feet (1 m) of shell-marl it must have persisted for a considerable time, and there may be a pointer to this in the Domesday fishery records for those villages on the Fen margin of the Breckland, that region of East Anglia made notably dry both by its porous soils and low rainfall. Yet Richard Fitzgilbert's Lakenheath estate is credited with four fisheries and a boat, Hockwold with six fisheries, and Methwold seven. Lakenheath village is within a mile of the corner of Red Mere, and Hockwold sits only 3 miles upstream near the Little Ouse river that formed its northern edge: these seem plausibly to be associated with the presence of the large mere and possibly so were the seven fisheries of Methwold also, although this village is somewhat further away. It is reasonable to suppose that the mere was drained by the cutting of the artificial channel of the Little Ouse from Crosswater Staunch to its present confluence with the Lynn Ouse at Brandon Creek near Southery, but of this again we have neither written historical nor separate archaeological evidence.

Unfortunately ploughing and drainage works have now often reduced the roddons in their upper reaches to a condition where the relationship to the meres is less easily recognised, and this is so for Whittlesey, Trundle, Ugg and Brick Meres (Fig. 30). However, equivalent evidence is not lacking. The diminishing roddon of the River Nene can be traced upstream from Ramsey, but as its banks lessen and disappear it is continued into the ancient waterway mapped as the 'River Nene' with meanders of the same magnitude, and with the meres we have mentioned all lying in the proper lateral relationship to this stream or its main tributaries. The ancient status of the waterway is confirmed by the way it carries county and parish boundaries that commonly date back a thousand years or so, and by historic references of about A.D. 1200 that strongly suggest the stream was an important means of transport almost as long ago. Written evidence also indicates that about A.D. 1020 the natural river channel, the 'Merelade', ran not through the mere, but along the winding route of a natural stream along its northern boundary. That in fact it had its own channel and banks of silt is further confirmed by borings made by the drainage authorities. A similar relationship exists between Trundle Mere and the winding channel of a stream beside it, now part of Yaxley Lode. There is thus every reason to regard these meres of the Nene system as also belonging to the primary lateral type of Red Mere.

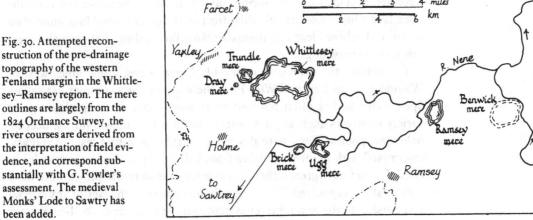

Fig. 30. Attempted reconstruction of the pre-drainage topography of the western Fenland margin in the Whittlesey–Ramsey region. The mere outlines are largely from the 1824 Ordnance Survey, the river courses are derived from the interpretation of field evidence, and correspond substantially with G. Fowler's assessment. The medieval Monks' Lode to Sawtry has been added.

Some few extinct meres, however, have a different nature. Jennings specifically compares Soham Mere with the 'side-valley' type of Norfolk Broad, created by the blockage of the valley exit during the submergence. Although the mere deposits at Soham are of silt and not shell-marl this has not been plausibly associated with its special mode of origin. The gentle topography of the Fenland margin has not generally favoured development of these side-valley meres. It is worth considering at this point how far we can recognise the conditions that allowed deposition of shell-marl in such large amounts, no doubt over considerable time, when at other stages of the Fenland history only thin transient layers of it have been found. It certainly pre-supposes fresh-water sources extremely rich in dissolved chalk, a need met by the extensive outcrops of Chalk, Jurassic limestones and big spread of chalky till in the catchwater areas of both minor and major streams entering the Fens. It seems requisite too that clear open water had to persist through the summer months, probably heating up considerably. It had also to remain free from silt or severe disturbance of lake-bottom muds that would quickly have smothered the algal bottom growth, and it had to remain over a long period too deep for invasion by the surrounding bulrush and *Phragmites* of the reed-swamp margins. The sloping roddon bank would to some extent help the second requirement, but it seems necessary to suppose that each mere represented some local concentration of the upland water supply constricted against the roddon on one side and by banks of deepening peat elsewhere. The constriction in the Whittlesey Mere and Trundle Mere region was apparently by the sloping banks of raised bogs already present before and during the transgression. This conjecture is elegantly supported by a letter written by the country poet, John Clare, in 1825 in which he mentions that Whittlesey 'is also a place very common for the cranberry that trails to the brink of the Mere'. Our only trailing cranberry in Britain is *Vaccinium oxycoccus*, a plant of acidic *Sphagnum* bogs: one could scarcely have better evidence of the former juxtaposition of the acidic bogs and the open lake. One can see exactly this effect on the great central limestone plain of Ireland where clear calcareous streams flow calmly between the great acidic *Sphagnum* bogs.

Far outside the main area of the black peat Fens, a few miles west of Wainfleet, in the East and West Fens there used to be a complex of shallow open water bodies often referred to as meres, but carrying also a great variety of names such as pitts, waters, lades, sikes, nukes, hurns, deepes, holes and pooles. They were described in 1789 as 'a vast tract of morass, intermixed with numbers of lakes from half a mile to two or three miles in circuit'. They gave great difficulties to successive drainage projects but have now totally disappeared. The variety in morphology, the lack of any special relationship to the natural river system and the absence of shell-marl from

Plate 26. Oblique air-view of Barton Broad, Norfolk. Its origin as a vast medieval peat excavation explains the remarkable extensions into and almost across the lake. They are the residue of the balks left between contiguous sets of diggings: the direction of the diggings changes at the parish boundaries. (Photograph by Department of Aerial Photography, University of Cambridge.)

them suggests that they differ in origin from the meres of the southern Fenland, possibly being no more than large peat cuttings such as the Norfolk Broads have now been shewn to be. The northern part of East Fen is shewn by Skertchly as peat fen with buried forest, but the peat was a small area constricted inside the foot of the Lincolnshire Wolds by the incursion of the estuarine silts of the Romano-British transgression that fully surround it to the east and south, inevitably impeding drainage of the fresh-water from the upland, and part of the mere system may indeed have rested upon the silt surface.

Some clue as to the original nature of parts at least of this tract of the northerly Fenland is given by the existence still near Friskney of 'Cranberry Farm' of which Alan Bloom wrote in 1953, 'cranberries grew in the unusually deep moss peat peculiar to the district'. This cannot have been other than a raised bog and to judge from the records that 'up to 4000 pecks were gathered in favourable seasons', it is unlikely to have been less in extent than half a square mile. At Hockham Mere in the Breckland and at Amberley Wild Brooks behind the Downs in West Sussex we have also the persistence of the name 'Cranberry Farm' but with additional direct evidence of the former raised bog in the form of residual peat of the characteristic type, a persistent living flora and sub-fossil pollen.

In the area of the south-eastern peat Fens diffuse and minor spreads of shell-marl are especially abundant and mottle big areas of the air photographs, but seldom or never offer opportunity for dating. However, in Wicken Fen, where the peat surface still remains as high as +7 ft O.D., the white shell-marl forms a fairly continuous layer 1 or 2 ft in thickness, covered only by a similar thickness of fen-peat. The pollen diagrams that pass through this deposit naturally confirm that it formed at a late stage in the sequence of British forest history, indeed at the stage when beech and hornbeam had for the first time become important woodland components in south-eastern Britain. This is very strongly confirmed in the pollen diagram from Wilton Bridge which discloses such high frequencies for these trees that one has to suppose natural beech woods were then growing on the chalk soils of the Fen margin. All the same the pollen analyses give no close evidence to date the marl, that still might be determined perhaps by radiocarbon dates of suitable samples of contemporary and uncontaminated wood found in it.

A separate occurrence of shell-marl outside the mere system, although very probably of the same age, was reported by Fowler in the shape of the former waterways that he called 'Old Slades'. His most striking example is now a bank of shell-marl that runs parallel to Reach Lode from the upland as far as the confluence with Burwell Lode and in an ill-defined way somewhat further towards Upware. From its straightness it has to be artificial and it had long been considered to be a Romano-British causeway from the frequency with which objects of this period had been found in it (Fig. 31). Fowler correctly identified the shell-marl as an open-water deposit and suggested, no doubt rightly, that this was an artificial Roman canal now left as a raised bank by the differential wastage of peat around it and by the fact that it had been ignored when peat-diggers cut ground above and near it. The pollen diagram made through it shewed the same undisturbed stratigraphy we had already encountered below the shell-marl at Wicken. Once again we have no clear evidence of when the marl formation ceased, but it is interesting that at the time when these investigations were being made in the 1930s, typical *Chara*-marl was still being thickly formed on the bottom of quieter stretches of Wicken Lode, and big bristly handsful of *Chara hispida* could be hauled to the surface with the release of the typical sulphurous smell of the bottom mud. Now that boat traffic so greatly disturbs the Lode this growth is no longer seen.

The Fenland meres have a particular advantage over all the earlier geological structures in that they alone persisted naturally in operation right

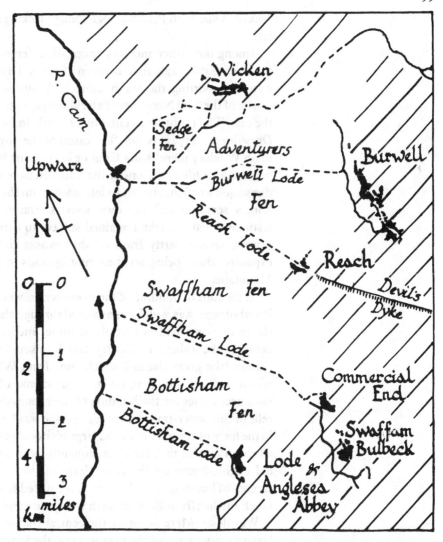

Fig. 31. Sketch map to show the linkage by lodes that are mainly artificial, of the eastern Fen-margin villages with the main river system, here the River Cam that, a little to the north, joins the River Great Ouse. Note especially the use made of the promontory at Reach where the Anglo-Saxon dyke and the Roman canal meet from the landward and Fenland sides respectively. Natural streams from the Chalk maintain water supply to the lodes, each of which terminated in a village quay or staithe to handle the boat traffic.

into the recent historic past. They commanded moreover such widespread popular interest that written accounts of them in relation to transport, rural economy, recreation and natural history extend over many centuries prior to their general drainage in the nineteenth century. It is not surprising that in so featureless a landscape as the Fenland, visitors and locals alike were drawn to make these great sheets of open water the centre of their surveys and holidays. It is natural too that Whittlesey Mere, with its impressive size of some 2 square miles, should among them have always commanded prior attention, a fact only distantly associated with the considerable historic importance of Whittlesey on its island of gravel-capped clay, 5 miles to the north east of the lake, a township that, in the early eighteenth century, ranked in size with Peterborough though lacking the cathedral status of that city.

Among the earlier and very enigmatic references involving the Mere are those of Saxon age that concern 'Cnut's Drain' or 'King's Delph', a waterway skirting the south shore of Whittlesey Island and linking the course of the 'Old Nene' near Peterborough with the natural river systems to the east. The tradition as cited by Dugdale in his *History of Imbanking and Drayning* . . . (1722) is that 'by reason of the boysterousnesse of the waves upon Witelsey mere' King Cnut or his consort had 'caused the ditch to be first made'. Other versions of the tradition have it that its digging followed disastrous wrecking during a violent storm on the mere of ships carrying the King's servants and his own sons. Certainly there is abundant later testimony that the lake remained subject to extremely sudden and severe storms, arising partly from its shallowness and partly from its extreme exposure, there being no sheltering uplands to the east short of the Ural Mountains.

At the time of building of the monasteries, when stone from Barnack near Peterborough was widely transported through the Fens, it is natural to find the mere mentioned as part of the route, and an agreement made in 1192 between the abbeys of Sawtry and Ramsey specifically exempted from closure 'the great channel which runs from Whittlesea Mere to Sawtry, which shall remain open, for by it the monks of Sawtry bring stones and such necessaries for the building of their monastery and of their offices'. A relic of this water-traffic in building stone was found on the bed of the mere in the form of several rough and large blocks of masonry, doubtless lost from a boat in transit: they have subsequently been moved to the vicinity of the old engine-house on the main drain of the reclaimed lake. Similar lost cargoes of building stone have been recovered from fenland waterways very much further from Barnack such as Upware and Prickwillow.

Whittlesey Mere occupies the central role in the odd publication *Lord Orford's voyage round the Fens in 1774*, the journals of two companions on the voyage, both of whom evidently lent themselves to the whimsy of his eccentric Lordship who organised a party of small sailing vessels, two or three dinghies and a towing horse 'Hippopotamus', to cross the Fens from Brandon for a sailing holiday on the great mere. The fancy they maintained was to describe the expedition, vessels, crew, navigation and all, in strictly nautical terms, but despite this general jocularity, the account incidentally gives much useful information on the local scene. Thus much of the 'Admiral's' time was taken with setting and visiting the trimmers put out in the lake, all the fleet indeed spent much time fishing and dined well on their catches. They report too on the local fishing rights and the indigenous anglers. The fleet occupied itself also with sailing trials between the boats, and experienced the dangers of one of the lake's sudden storms. Still with the air of a naval expedition reporting upon hitherto unexplored regions, the journals report upon the topography, vegetation, customs and aspects of the aboriginals, though they are respectful regarding dining with the Bishop of Peterborough!

It is apparent from pre-drainage maps that by the late eighteenth century the local gentry had their own boat houses at suitable points on the mereside, and regattas were great summer occasions for all classes of the local population. These summer jollifications were, however, outdone by those following upon long hard winters when the lake surface was frozen, and a large part of the rural population kept from work on the iron-hard land, repaired with their locally made 'Whittlesey pattens', sledges and such warm clothing as they had, to enjoy the winter sunshine and see the skating matches that were and have remained so characteristic a feature of the Fenland. In fact the great traditions of Fenland skating owed less to the great meres than to the long straight stretches of shallow quickly-frozen flood-water in the 'washes' beside the big fen drains, and the techniques of long-distance and speed running were fostered by these abundant facilities.

By the early nineteenth century the progress of drainage in the surrounding Fenland was having such effect that intermittently the mere was often, if only temporarily, three parts dry and it needed only the continued improvement of mechanical engineering and the vision and drive of a local landowner such as William Wells to initiate and carry through the total reclamation of the mere. It will be appropriate, however, to consider this final phase as part of the whole story of Fenland drainage and its consequences (Chapters 13 and 14).

II

Conspectus and historical framework

Now that we have surveyed the evidence that the quest has made available for the past millenia it may be advantageous to abandon close-range enquiry for the moment, and attempt to survey in very general terms what has happened to the Fenland through this long time. We may perhaps imagine ourselves as long-lived geographical observers able to watch from a satellite viewpoint the speeded-up progress of change in our area. We can be guided by the schematic section of Fen deposits (Fig. 32) and by a later reconstruction of the date and vertical extent of changes of Fenland sea- and land-level (Fig. 33).

It was as recently as 20 000 years ago that ice sheets stood on the north shore of the Humber and it is not unlikely that this was also the time when the last advance of the glaciers reached the Norfolk coast at Hunstanton. It is supposed that this North Sea ice may have ponded back melt water to form a large Fenland lake whose outflows were across the low watershed to the east, along routes such as the Waveney valley, but evidence for this is extremely scanty.

By the time the ice had gone and the Flandrian period was established, around 7500 B.C., the area we now call the Fenland would have had no claim to such a title. The floor of the shallow basin revealed the wide parallel outcrops of Jurassic clays, Greensand and Gault clay, heavily plastered in some parts with Boulder Clays or glacial gravels. That portion between Littleport, Mildenhall and Brandon was covered by a Boulder Clay influenced largely by movement of the local ice over the local Greensand, and it was no doubt already heavily leached, loose and acidic in character. In this Pre-boreal time through all this countryside there probably extended open woodland of pine and birch, resembling in general character that of northern Europe or northern Canada today. Only locally favoured spots carried the trees of warmer climates. Through this sparsely wooded region ran the wide river valleys, cut when the volume and load of their waters were much greater, and now mostly filled with flat sedge fen bordering a shrunken stream. Some two or three hundred miles downstream these rivers merged into the Fenland of that date, and over the bed of the present North

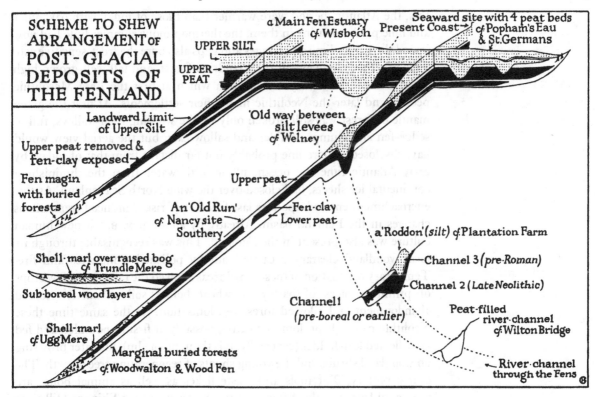

SCHEME TO SHEW ARRANGEMENT OF POST-GLACIAL DEPOSITS OF THE FENLAND

UPPER SILT

UPPER PEAT

a Main Fen Estuary of Wisbech

Seaward site with 4 peat beds of Popham's Eau & St. Germans

Present Coast

Landward Limit of Upper Silt

'Old way' between silt levées of Welney

Upper peat removed & fen-clay exposed

Fen magin with buried forests

Upper peat

An 'Old Run' of Nancy site Southery

Fen-clay Lower peat

a 'Roddon' (*silt*) of Plantation Farm

Channel 3 (*pre-Roman*)

Channel 2 (*Late Neolithic*)

Shell-marl over raised bog of Trundle Mere

Sub-boreal wood layer

Channel 1 (*pre-boreal or earlier*)

Peat-filled river-channel of Wilton Bridge

Shell-marl of Ugg Mere

Marginal buried forests of Woodwalton & Wood Fen

River-channel through the Fens

Fig. 32. Schematic solid section to show the type of variation in Fenland deposits at different levels and different distances from the sea to the Fen margins. First published in 1940, the disposition has been generally substantiated since by various age determinations of the deposits and specifically by the radiocarbon ages cited in the text (Chapters 4 to 11).

Sea, Mesolithic hunters and fisher-folk could reach the rest of the Continent: the sea then lay north of the Dogger Bank.

As the climate now continued its decisive improvement in the Boreal period, warmth-loving trees replaced the birch and pine, more rapidly on the clay soils than on the sands. Gradually through the course of a few centuries, the landscape saw replacement of birch by pine, and the establishment of elm and oak woods with hazel both as undergrowth and as hazel scrub. No doubt a generous cover of deciduous forest soon clothed all our present Fenland save the wet water-courses and infrequent lakes. The increasing geniality of climate towards the end of the Boreal period allowed the alder and the linden to appear in locally favoured places. Mesolithic man was now encamped at Shippea Hill on the sandy river banks bordering the original Little Ouse and the valley fen was strewn with bones of animals killed in the forest and with microlithic flints from his weapons and tools.

At around 5500 B.C., the Fenland forests became affected by the increasing wetness of the climate, and the alder everywhere assumed a rôle of great importance in the wetter woodland glades frequent on the clay soils and along stream and lake margins: this rôle it never relinquished until the historic clearance and drainage of English forest land. The woodlands of this, the Atlantic, period were warmer than those of today with oak, alder and lime predominating in them: the thermophilous summer lime had now joined the winter lime and fringes and gaps also bore the frost-sensitive ash. This was the golden age of the Flandrian forest, and tall well-grown high forest covered a countryside through which at first the Late Mesolithic people, and later the Neolithic made their settlements. At first the green mantle of forest was still broken only by the winding river-valleys, full of sedge-fen and clumps of alder and sallow carr, but the aerial view would have disclosed a shore line probably not far distant from that of today: by early Atlantic time the ocean, filling with water from the diminishing continental ice sheets, had closed over the wide North Sea fenland. As this encroachment entered upon its last few feet of rise it induced tremendous changes in the Fenland basin. Already by about 3000 B.C. a new human culture was also present in the Fenland. This was recognisable through its local woodland clearances carried out by polished stone axes and fire. Temporary occupation of these small areas allowed the raising of a few crops of primitive races of barley or wheat before soil exhaustion forced abandonment and allowed forest recolonisation. At the same time these Neolithic people kept domestic cattle, possibly at first impounded and fed on collected leaf-fodder (especially, it is thought, of elm), but later browsing on woodland shrubs and the young shoots of regenerating tree growth. The coarse pottery, flint tools and waste flakes as well as animal bones are recovered from people of this culture who occupied the Shippea Hill river banks long after the Mesolithic hunter-fisher people had vacated them. The highly characteristic polished stone axes of the Neolithic culture, some traded into the Fens from far afield, were lost here and there on the forest floor, to be revealed only by the peat-diggers' beckets or the spades of the coprolite workers when they exploited the south-eastern Fen margin in the nineteenth century.

The inexorable rise of ocean level was now inducing a general waterlogging of the whole Fenland basin and even in those landward parts of it into which sea water did not reach, the fresh-water from the uplands was ponded back without escape seawards. Thus with the growth of reeds and sedges, black fen peat began by about 2500 B.C., to cover the woodland floor, and to embed the huge trunks and stools of the forest trees killed by exclusion of air from their roots. Only rapidly growing soft peat could have sealed up these great high forest trees before decay could remove them. Thus the landward parts of the Fenland became converted to a vast tract of sedge-fen constituting in effect the area of the black peat land of the present day. Open water was rather infrequent and the sedge-fen was naturally subject to the progressive changes of vegetational succession so that it was progressively invaded by alder, sallow and birch carr that even, in the shallower margins, progressed into fen woods with oak, pine and yew. At best, however, the Fens remained extremely inhospitable and little frequented by Neolithic man.

At the height of the marine transgression, between about 2500 and 2000 B.C., almost the whole of the Fenland area had become a vast brackish lagoon a few feet in depth, where fresh-water from the uplands was held up by the high mean level of the sea and was mixed with the salty incursions of spring tides. It is likely that coastal banks of silt and sand were formed, and behind them in this large shallow lake was deposited all the silt and clay brought up from the sea by the invading high tides. The waters supported a limited range of brackish-water organisms, mostly microscopic, although truly marine creatures, even such as whales, strayed into them from the sea now and then. At the land–water edge the water was fresh enough to support a dense growth of the common reed, much as it now grows on the shores of the Baltic. Up the river channels these reed beds gave place to fresh-water sedge-fen, and elsewhere round the Fen margin the brackish water must have extended right up to the existing fen woods, coming indeed close to the upland Fen boundary itself. Within a few hundred years many feet of soft Fen Clay had been laid down over the earlier fen peat. We do not know the whereabouts of the main estuary of the Fenland at this time, but it may well have been north of the historic estuary at Wisbech.

In sharp contrast to this régime of literally overwhelming wetness, the Fenland basin from about 2000 B.C. experienced a period of remarkable relative dryness, due in part possibly to the fact that this was now the Sub-boreal climatic period generally considered as essentially warm and dry, but more certainly to a standstill or regression of sea-level. This led to the re-establishment of fresh-water fen over a large part of the Fenland basin, clothing the flat surface of the Fen Clay with sedge-fen interrupted by local shallow pools of open water, but all developing naturally through the fen communities of alder, birch, willow carr towards fen woods, and soon mantling the clay surface with an increasing thickness of peat. This process began at the Fen margins and extended progressively seawards, and in the more marginal parts the fen woods that had been hardly affected by the marine invasion continued to form woods of pine, yew and oak. In the rather isolated basin between Holme and Yaxley likewise, acidic raised bogs continued to grow in conditions where flooding by alkaline water from the uplands had become much less frequent. Similarly the marginal fen woods here and there shewed signs of becoming acidic, and the overall dryness even allowed raised bog to form as far seaward as Nordelph.

There can be little doubt that prehistoric man must have passed freely over these dry fen surfaces and through their fen woods, for the weapons and tools of Bronze Age man occur very abundantly through the Upper Peat

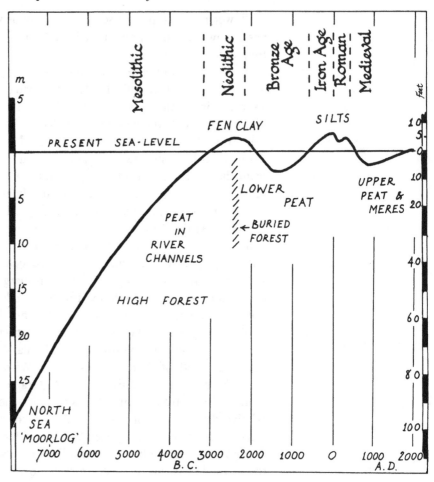

Fig. 33. Changes in relative land- and sea-level in the English Fenland through the Flandrian Period (with the present as zero), together with the main shifts in Fenland conditions and deposits that they induced. The great initial rise in sea-level was due to the melting of the ice sheets at the close of the last Glacial stage. The Fenland succession of human cultures is broadly shewn also.

and extend from the upland margin some miles out into the peat fen itself. At Shippea Hill the river banks were settled by Early Bronze Age people soon after the end of the marine transgression and there have been widespread isolated finds of the same period. In the *Middle* Bronze Age, larger woodland clearances than those previously known led to much larger spreads of agricultural weeds in the local pollen rain and to thin clay layers washed into the raised bog at Holme Fen around 1440 B.C., from nearby uplands. It was, however, in the *Late* Bronze Age that the Fens were especially heavily occupied: not only do the traces remain as abundant individual finds but it is evident that passage about the Fens was facilitated by the construction of wooden trackways, whilst the burial of bronze-founder's and other hoards in the peat, with the implied intention of revisiting the site, indicates a fair degree of continuing accessibility. One presumes that at this period the coast had retreated far towards that of the present day but it is hard to come by exact evidence of its position, so much was destroyed or buried during the succeeding phase of Fenland history: fresh-water fens, however, certainly occurred far seaward of their present limits.

In the middle of the last millenium before Christ the Fenland entered its last great natural reversal of wetness. The climate now became (from the human standpoint) much worse, with the opening of the colder and wetter Sub-atlantic climatic period: the warmth-demanding lime tree practically vanished from the forest cover of the upland although it is true that exploitation by Iron Age man may have had something to do with this. This climatic deterioration was combined with the onset of a second phase of extensive marine transgression that was to reach its culmination in the Romano-British period, and restore extensive waterlogging throughout the peat fens. In striking contrast to their previously dense occupation, now the Fens were entirely shunned by Iron Age man, and it is safe to assume that widespread open sedge-fen was re-created above the dry Bronze Age fen woods.

The phase of marine invasion that now began, differed from the earlier one in that it did not produce a large inland lake to be filled with brackish mud, but gave rise to extensive coastal salt-marshes of silt and fine sand: these extended inland along a great estuary centred where Wisbech now stands. There it was many miles wide but it constricted gradually upstream towards Welney whence the tidal silt extended up the meandering river channels, forming raised banks or levées that reached far into the peat country, diminishing in height progressively as the tidal range in the channel lessened with increasing distance from the sea.

Although there is little decisive evidence, it seems likely that during the pre-Roman Iron Age there built up round the Wash the greater part of the thick deposits of tidal silt that constitute the fertile 'Marshland' of the Fens, and that offered its surface to the colonisation of Romano-British settlement from the mid first century A.D. This occupation was at its densest at the close of the second century, when farms must have been at least as numerous as they are today, but then a slight renewal of marine transgression with a thin deposition of fresh silt prevented occupation until it resumed in less intensity than before, in the first half of the fourth century.

At its height, the very considerable agricultural settlement of the salt-marsh surfaces was associated with a multitude of minor engineering works of drainage and water-transport that have left the landscape decorated by an intricate pattern of small farmstead fields, droves, drains and persisting marsh creeks, all intriguingly visible from the air. The raised banks of the tidal streams that meandered through the peat Fens likewise attracted Roman-British occupants and well-engineered transport canals, usually straight, were constructed not only marginally but also more centrally and in relation to the main river systems.

The building up of the tidal silt levées of the rivers, now reaching its maximum, had the effect near the limit of tidal penetration of blocking off escape seawards of the fresh-water that continued to enter the Fens from the upland, so that it accumulated as a series of large shallow lakes close alongside the raised river banks. Thus originated the great Fenland meres such as those of Whittlesey, Ramsey and Streatham, most of which were to persist until the era of steam-assisted drainage of the Fens. The general waterlogging also caused similar shallow hard-water lakes to form more widely throughout the peat landscape, but these were more transient, succumbing to the processes of colonisation and infilling by fen vegetation that always tend to reconstitute a general mantle of fen peat. This in general terms was the condition of the Fenland at the end of the Roman occupation: despite colonisation of the silt land the peat hinterland had throughout been unoccupied and it remained largely so until the drainage of the seventeenth century. Throughout the dark ages the Fenland was virtually untenanted: not only do the wet lands yield no trace of human occupation but even the fertile silt-lands exploited in the Roman period were abandoned. We had indeed attained that condition quoted from the life of St Guthlac, founder of Crowland Abbey, of a 'hideous fen of a huge bigness which, beginning at the banks of the river Gronte, extends itself from the south to the north, even to the sea', and according to legend the home only of fenmen or devils that, real or imaginary were a source of great torment to the saint.

This highly condensed account hinges substantially on the investigations made by the Fenland Research Committee. Although fragmented by the Second World War and never effectively reconstituted afterwards it had nevertheless put together a plausible broad outline of Fenland history representing a synthesis it would have been hard to attain except by some such concerted attack from many angles and with several kinds of expertise. So far its conclusions have held up astonishingly well, not least when radiocarbon dating became available to give an absolute chronology to the sequences of events and cultures it concerned.

There remain unanswered of course a substantial number of questions. Reflection upon the two major episodes of marine invasion that have been registered in the Fenland history, the Fen Clay incursion and that of the Upper Silts, returns constantly to the question of why the two are so different in the nature of their sediments. The Fen Clay was exceedingly widespread, coming close to the Fen margins, it was lagoonal and, apart from its onset, largely brackish in character, whilst in its later stages its flat surface was reticulated by a fine mesh of salt-marsh creeks feeding into the main river channels. By contrast the Upper Silts consisted of a wide belt a few miles deep around the Wash, extending as a tapering main estuary whose chief tributaries built raised levées but deposited no general mantle of silt or clay behind them: these Upper Silts were the product of quite saline conditions. The contrast is all the stronger since the later, Romano-British tidal streams occupied the same channels that had been used by those of the earlier episode.

Perhaps we can best explain it by recollecting that the Fen Clay incursion marked only the last stages of the great eustatic rise of ocean level. The final few metres' rise in restoration of the North Sea, after submerging the Dogger Bank, would have encountered a barrier of glacial drift such as the Hunstanton Boulder Clay episode may have left across the mouth of the Wash when its glaciers had earlier come southwards from Lincolnshire. When any such barrier was breached by the rising sea one imagines a quickly widening break-through and widespread inundation of the shallow Fenland basin lying behind. On the other hand, after recession of the sea at the close of the Fen Clay episode there had followed at least a thousand years of fresh-water conditions during which a mantle of fens and acid bogs had built up peat over the whole extent of the Fen Clay surface, or at least far seaward of its present limit at the edge of the coastal marshland. The peats can scarcely have been less than 6 to 9 ft (2–3 m) thick and deeply concealed the minor creek systems at the top of the Fen Clay, ceasing only at the banks of the deeper fresh-water rivers passing seaward through the Fens from the uplands. When the Iron Age/Romano-British marine invasion took place it was into a landscape of this kind and through a stable coast that had reformed during a millenium: outside this coast the Wash was full of sand banks and shoals much as it is today. It seems likely that it was these circumstances that restricted the access of sea-water to the high tides carrying their burden of silts, that penetrated only the main estuary and a little way into its main distributaries.

It has been a consistent feature through the evolution of the Fenland that the main river beds served equally for all the successive phases of it. Cut during the later part of the last glacial stage, as we saw at Shippea Hill, it was within the same bed that the Fen Clay main drainage creeks were active and again it was here that the big roddons were built during Romano-British time. Quite possibly the continuing passage of fresh-water rivers from outside the Fens guaranteed their persistence even through episodes of marine invasion. This persistence would explain the odd effect presented in recent air photographs, that show the minor creek systems of the top of the Fen Clay all converging into these same river channels. This effect, only become visible now that the Upper Peat has substantially wasted away, appears at first glance to indicate that the fine creeks and the roddons are contemporary. This is the trap of the palimpsest, well known to interpreters of air photographs, the superposition on a parchment of younger writing above an older one. In fact it seems clear that there was no salt-marsh creek system behind the Roman roddons, and that nearly two millenia separate the two phenomena that lie with such deceptive congruence together in the aerial views. This has to be particularly borne in mind in seeking to interpret the maps of surface Fen deposits so recently made by the Soil Survey.

12

Peat and its winning

A large part of the population of the British Isles has only a very hazy idea, and that based upon the seedsman's bags of 'horticultural peat', of the meaning of 'peat' and, for that matter, it is a term unfamiliar to the fenmen who till it, dyke it and in the past have cut it for fuel: to them it has always been 'turf'. It is a very broad category of substances that, however, possess certain things in common: they arise invariably by the accumulation of plant remains *in situ*, i.e., directly from the plants growing on the peat surface. This accumulation is the consequence of the arrest or slowing down of decay, primarily because of lack of oxygen beneath the surface in any more-or-less permanently waterlogged deposit. In consequence, the peat is almost totally organic, with absence of all sand, silt, clay or ash except that fortuitously introduced by events that have nothing to do with growth of the peat itself, and this of course explains its effectiveness as a fuel, releasing, when burned, the solar energy fixed initially by the green foliage of plants during photosynthesis. So long as waterlogging persists the growth of a peat deposit will continue, so that considerable thicknesses may be formed, certainly as much as 30 ft (9 m) and in the lower layers of such mires the peat is subject to great compaction as well as slow processes of chemical and bacterial alteration still little understood. These more altered peat layers come to resemble in many ways brown coals and lignites like those exploited in the Ruhr and in the coal fields of Australia. Indeed it is virtually certain that these fuels originated as peat, and it is likely that in far more distant geological time, so did a good deal of the more valuable hard coals.

From what has been written from Chapter 2 onwards it will have become apparent that despite the over-riding common properties, peat is likely to exist in a very wide range of types that differ according to the circumstances of the waterlogging that induce mire-formation (thus fen, raised bog and blanket bog), and to the stage of vegetational succession in which any particular peat layer is formed (thus for example, reed-swamp, sedge-fen, brush-wood, fen oakwood or *Sphagnum*-bog). To these must be added the varying degrees of compaction, and of the biochemical alteration and progressive destruction of plant material that is generally referred to as

Plate 27. Fibrous aquatic peat rich in the fine roots and rhizomes clothed with loose papery leaf-sheaths, characteristic of the sedge-like *Scheuchzeria palustris*. This plant, entirely typical of open pool margins on acid bogs is virtually extinct in the British Isles but its remains have been found in many scattered raised-bogs among them Holme Fen, Hunts.

Plate 28. *Cladium, Hypnum*-moss peat. The large red rhizomes of the sword sedge have been cut across transversely.

humification and that, especially in the acidic peats, leads to the development of a strongly colloidal texture, associated with great powers of water-retention. These humic colloids that are especially valuable horticulturally, turn upon drying, into a hard horny substance that doesn't resume its jelly-like character upon rewetting: it makes, however, the turves of the old humified 'black' peat quite reasonably hard and durable.

It is not difficult to see that the most natural and most informative way of classifying peats is to refer them to the ecological conditions in which they have originated, and we have in effect already shewn how this can be readily achieved in the Fenland by identifying the still recognisable plant remains in the peat, be they of the dimensions of 'bog oaks' or of microscopic pollen grains falling upon the growing surface of the mire. Some of the identifications can be made in the field and some experienced research workers have remarkable ability to recognise small fruits, leaf and stem fragments, even part decayed, that they recover from their peat borings, so that as the day's work progresses they form a mental picture of the past vegetation that, layer by layer, formed the deposit they are working upon. Such recognition is naturally made much closer and more dependable when samples are taken back to the laboratory, so that after maceration in dilute acid or alkali that dissolves the humic matrix, the small plant fragments can be sieved off, washed and compared, using the microscope where necessary, with standard collections of named material.

We have already seen what a great gain in comprehension of the evolution of the Fenland is obtained by recognising this range of peat types and their vast sub-fossil content. Although much of this gain is scientific, we are also helped by it to some extent in explaining the former economic exploitation of peat. The Fenland seems, at least on first consideration, to be somewhat at variance with the general rule, for the vast majority of deposits now cut, commercially or domestically, are in acidic raised bogs or blanket bogs. To some extent no doubt this reflects the general sparsity of alternative fuel in the regions of the West and North, or on the high mountain areas where such mires are prevalent, but in addition they provide a type of fuel that is easier to cut, freer of timber and that burns better and with far less ash than the peats derived from the calcareous fens. This applies especially to the highly humified, dark-coloured *Sphagnum* peat with cotton-grass and heather that constitutes the lower and older part of raised bogs, and that is commonly known as the 'old' or 'black' peat, to distinguish it from the younger and upper layers, often 2 or 3 metres in thickness that overlie it. This much paler peat, the 'young' or 'white' peat consists of relatively unhumified and uncompressed *Sphagnum* peat, still loose and spongy in texture and in which all the moss leaves and stems have retained their exquisite microstructure, in which the green chlorophyllous cells form a network separated by the large empty perforate cells that give this genus of moss its extraordinary powers of water absorption and retention. It is these powers that not only make this moss able to waterlog the ground and build up great domed mires, but which, in recent years, has made possible the sale of the fragmented 'white peat' in vast amounts for horticulture. As long ago as the Bronze Age, *Sphagnum* moss was being employed as a surgical dressing: a skeleton of this age found at Methil Hill, Fifeshire, shewed a severe chest wound still covered with a large swab of *Sphagnum*. This moss was extensively collected for dressings during the 1914–18 war and the fibrous pale peat has been also much employed therapeutically. The great adsorptive ability of the undecayed *Sphagnum* peat explains its long-standing employment as stable litter, and probably lay behind the First War experiment by a large commercial undertaking in the Fens of combining peat of this kind with molasses to make cattle food.

Of course, beneath the great majority of raised bogs lie the deposits of fen peat that reflect the derivation of so many of them from the infilling of lake-basins, but the peat-diggers of such sites generally cease cutting when these layers have been reached, dismissing them as troublesome to cut and the peat not worth the digging.

There is little doubt of the extreme age of the practice of peat winning, especially in areas of early forest clearance. In East Anglia, where such clearances had begun in Early Neolithic time and were active when Thetford was the thriving centre of the East Anglian Kingdom, it was nevertheless extremely startling to find ourselves faced with incontestable evidence that all the Norfolk Broads had essentially been created as peat-diggings during the later Middle Ages. One learned shortly afterwards that the Dutch had reached a similar conclusion that equally large lakes in Holland, now extensively used for sailing, had an exactly similar origin. What might be a comparable instance may have been formerly present in north-eastern Fenland. The extremely unordered arrangement and the odd and miscellaneous shapes of the meres that formerly existed there in the East Fen of South Holland seem to suggest no evident geological cause, and raise in one's mind the conjecture whether they may not indeed, like the Broads of Norfolk, have been created solely by intensive peat-cutting. The East Fen was actually the area of peat fen far most easily accessible to the silt lands in which tidal water was available, sitting just behind the narrow coastal belt on which stand Wainfleet, Wrangle and Friskney, all of them settlements for which there is specific medieval record of the importation of turves for their salt extraction pans. As Professor Darby has made clear in his *Medieval Fenland*, although there are no actual Domesday records of turbary evaluations in the Fens, there are numerous mentions of turbaries and turbary rights between about 1200 and 1400: many concern disputes

Plate *29*. Last stages of peat-cutting on Swaffham Fen about 1922. The shape of the turves shews that they were cut with a becket but the tool left beside the peat stack is a typical peat spade.

between religious houses, such as the recurrent quarrels between those of Crowland and Spalding over conflicting usage of cattle grazing on the peat-drying grounds of Deeping Fen. This reflects a large body of customs and usage governing turbary rights, such as times and location of cutting and protection of the susceptible rows and stacks of drying turves. Such a body of custom and law was of special importance in relation to the widespread common rights of turbary, rights the recollection of which has now almost totally gone from the working fenmen. However, the operation of such rights here and there ceased comparatively recently. At Wicken Fen there are two areas, one between Monk's Lode and Wicken Lode, known as Wicken Poor's Fen and the other in the main sedge-fen, referred to as Wicken Poor's Piece, each of which was available for peat extraction and sedge-cutting by the poor of Wicken after the third Monday in July each year, and here a man might take what crop or what peat he himself could get in the season, but he might not employ assistance. This probably explains the random scatter of separate small pits, arranged on no common plan, that still characterises both areas and most strongly contrasts with the very broad-scale co-ordinated plan of rectangular trenching that I suppose only to have become possible after the division of the fen following the Enclosure Act. As I recollect the Poor's Pieces in the 1920s the sides of many of the pits were still extremely steep, so that it cannot have been many years since they had last been dug, possibly ten or twenty.

It seems likely that similar local peat extraction took place along the

Plate *30*. Peat cart with a Burwell name plate, probably early 1900s. (Photograph by W. S. Farren.)

margins of all the southern peat Fens from medieval times, especially where there were adjacent religious houses or villages: the inaccessible central areas would be largely left uncut since local populations were small and transport of the bulky product was then so difficult. My guess would be that only comparatively late, when the effects of widespread drainage in and after the seventeenth century were reducing water-levels and allowing some transport by roads and dykes, and when ownership of land, following the awards of the early nineteenth century, had resulted in larger holdings, was there large-scale peat-extraction. This period too was doomed to a relatively short life since the better transport and expecially the advent of railways, allowed the easy introduction of the far more efficient hard coal, much of which travelled by sea-route and up the Fenland waterways to staithes at the head of the fringing lodes.

However, it is this period of larger-scale peat-digging that has left its traces on the landscape and in the memories of recent generations of fen people. Of course it is fairly certain that large-scale concerted peat diggings were made in such medieval projects as involved contracts to supply large quantities of turves, and such must have been involved to excavate large areas like Barton (Plate *29*) or Hoveton Broad. All in all it seems likely that the methods, tools and nomenclature of the turbaries, large or small, have behind them a long traditional usage. We may note that what is now called 'Wicken Sedge Fen', was referred to in a survey of 1684 as 'Wickin Sedg fenn and Turf fenn'.

For our knowledge of tools and their manner of use we again rely largely upon Skertchly, writing of the period when he was collecting field evidence for his *Geology of the Fenland*, that is presumably a decade or so before 1877. There are also a few less detailed later accounts, the best of which is the illustrated account by Ennion (1940) of the peat cutting he was familiar with on Burwell Fen just before the industry there ceased. Some ancient turbary tools are kept in the local museums. As we can gather from the persistent names of one or two Fenland pubs, in Chatteris and Littleport, the main tools were the 'Spade and becket'. Of these the becket is the tool most specifically developed for peat cutting and is substantially the same tool as that used in turbaries over almost the whole of the British Isles, corresponding to the Irish 'slane' or 'loy'. It is essentially a rectangular, perfectly flat, spade usually of wood shod at the bottom edge with iron, and characteristically carrying at the foot of the blade a short triangular iron flange whose lower cutting edge thus continues that of the blade itself at

Plate *31*. Aerial view of a substantial raised bog at Whixall on the Denbigh–Salop border. The parallel peat-trenches extend over most of the bog except the uncut far margin next the agricultural land on the Boulder Clay. This conveys something of the speed with which large areas of peat can be cut away. (Photograph by Department of Aerial Photography, University of Cambridge.)

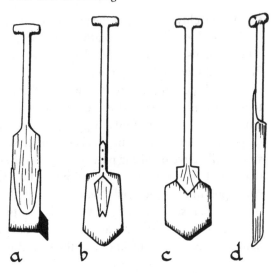

Fig. 34. Peat-cutting tools of the Fenland; *a*, becket; *b* and *c*, turf spades; *d*, turf knife. The first three are essentially wooden-bladed tools shod with iron and they existed in some variety: there were also shovels or scoops entirely made of wood. Note particularly the sharp basal flange projecting forward from the base of the blade of the becket.

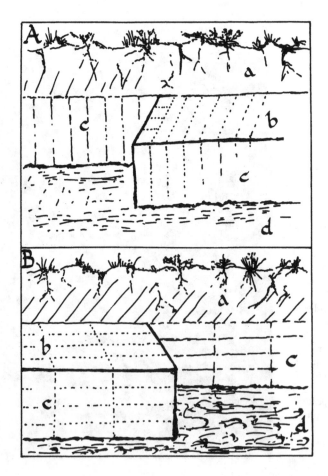

Fig. 35. Modes of turf-cutting with the becket, alternatively with vertical cuts from above (A), or horizontally from the side (B), where the peat is fibrous and strongly layered. In either method the loose upper 'hoddy' peat (*a*) full of roots and often cracked, is roughly cut away with a peat spade and tossed into an existing peat trench: the surface of the good peat (*b*) is smoothed and serves as a working platform. The dotted lines on the uncut platform shew the position of the next cesses to be cut, those in A by the cutter standing on the platform, those in B by him standing on the excavated floor of the trench (*d*). The side wall of the trench commonly shows marks of recent peat cutting (*c*) made by the becket and turf knife.

right angles (Fig. 34). Thus when the spade is driven vertically downwards just behind the open peat face, the one stroke makes two cuts together and the rectangular 'cesses' of turf (cut in succession from the face) can be lifted on the blade of the becket to be deposited on the level surface of the balk of still-uncut peat. The effective use of the becket depends upon having a coherent, rather amorphous type of peat, so that in opening any fresh peat trench the first step is to cut through the weathered upper peat penetrated by living roots, rodent burrows and cracks. This 'hoddy' peat is disposed of by the extremely sharp and well-polished peat spade, again quite flat and with the shaft and blade in precisely the same plane. This rough peat is often thrown into the wet floor of the trench from which peat has already been cut, so initiating the formation of a new surface for use as pasture or drying ground. Alternatively the great irregular lumps may be sold, though at a low price since they are a poor quickly-consumed and dirty fuel. The hoddy peat having been removed along the length of the cutting, a long turf knife or the flat spade is employed to prepare a meticulously level and even working surface at the top of the good peat, and a beginning having been made at one end of the trench, the digger thereafter stands on the prepared surface and, vertical cuts at one side and end of the prospective trench having been made with the turf knife, he uses the becket by cutting from that end, row after row of cesses, each row containing the precise number the width of trench has been designed for (Fig. 35). The diggers take pride in the exactitude of their work, and the absence of spoiled turves or broken edges is very pronounced for all that the work is exceedingly hard manual labour. As Skertchly reports, the size of becket varied greatly from one district to another and correspondingly the cesses differed in size: he notes a variation from $20 \times 5 \times 4$ in to $12 \times 3 \times 2$ in at one place, Isleham, but cites as average those of Coveney at $9\frac{1}{2} \times 6\frac{1}{4} \times 4$. They all agree in the greater length for the dimension of the downward cut. He indicates that a good digger in a fourteen-hour stint would cut eight to ten thousand cesses. Depending upon the local conditions, especially those of the peat deposit, demand and local drainage, from the one trench cesses might be cut one, two or even three rows deep: the labour of lifting the water-saturated blocks of peat from the bottom layer to the uncut surface was of course very much increased. The peat trenches in the Fens were usually about a yard wide (1 m) separated from one another by balks of uncut peat initially 5 or 6 yards wide (*c.* 5 m). With extended digging, the trenches were progressively widened at the expense of the balks, drying of turves being transferred gradually to the newly consolidated floor of the previously cut trench. Eventually the last balks were cut away and the whole area became free either for use as pasture or perhaps at a later date, for a second cut into the deeper layers of peat. Sybil Marshall's father recollected that when he, as a young man (that is,

Plate *32*. Peat cutting of the extensive raised bog of northern Holland makes use of water-transport through the wide Staatscanal. Many of the Fenland waterways, large and small, no doubt served a similar purpose though loading was no doubt more laborious. Klazinaveen, 1959.

about 1880), helped in cutting part of Woodwalton Fen, two rows of cesses each 14 in deep were cut at a time, and that within the two years' lease this part of the Fen was cut twice over, that is to say a total depth of 4 ft 8 in (*c.* 1.4 m), excluding the 'hoddy peat'. This he calculated would yield the digger some 400 000 turves per acre. 'Will 'En' makes the important point that the exploitation of 'Jackson's Fen' was preceded by the cutting of parallel boat dykes by means of which the peat was carried by barges into the Raveley Drain and thence to quite distant markets. This co-ordination with water-carriage no doubt in part explains the many residual dykes on Wicken Sedge Fen and certainly water-transport still plays a big rôle in peat

Peat *33*. Peat digging of the drained raised bog of Shapwick Heath, Somerset (*c.* 1950). On the left foreground the (spade-cut) cubes of peat, that in the more distant part have each been sliced into three along the growth planes. On the right two stages of arrangement of the turves for drying before transfer to big stacks on the droves.

extraction in the northern Netherlands despite progressive take-over by the lorry (see Plate *32*).

Skertchly records that when the 'good' peat was liable to break on the becket, as when it is strongly fibrous or horizontally bedded, another technique of cutting was employed. The flat clean working surface having been exposed as before, it was carefully laid out with a sharp marker and along these lines precisely vertical cuts with the flat spade were made, so isolating rectangular peat blocks: these were then cut *horizontally* along the bedding planes by a specially wide becket. In the Somerset Levels, the only area where I have seen this method employed, there seemed in fact to be no other and there the whole cutting operation was performed with the peat-spade, the becket apparently not in use at all. The peat blocks about 10 in (25 cm) cube, and weighing about 18 lb (8.2 kg), were lifted intact and placed on their sides on the drying surface to be arranged by an assistant with a type of hay-fork, prior to a neat splitting with the spade along the bedding plane into three equal turves (Plate *33*). It was in the Somerset Levels that I saw the peat-diggers employing a very long-handled scoop slung from a gallows, to eject the accumulated water each morning from the peat trench before recommencing digging (Plate *34*). This hard manual labour was evidently the precise equivalent to that used by the Fenland dykers who spoke of 'lecking-out' the drain, and it seems highly likely that

Plate *34*. A long-handled drainage scoop slung from an improvised gallows, the scoop adapted from a metal tin. To rid the working trench of the overnight accumulation of water ('laving' it out) the digger stands on the plank and swings the scoop down into the trench, up and over the low bank so that a twist discharges the load into the dyke beyond. This device is no doubt the equivalent of the 'lecking scoop' of the Fenland. Shapwick Heath, Somerset, 1947.

the wooden 'leck' was similarly employed in the Fen turbaries. It was not unlike the medieval 'dydal', a similar scoop for getting peat from below water-level, an instrument the use of which was sometimes explicitly forbidden in tenancy agreements.

By whatever method cut, the turves have next to be dried. After a day or two to give them a little strength they are arranged in 'windrows' or small open piles of ingenious architecture with large spaces for wind movement between the drying blocks of peat. The turves may be rearranged to dry all surfaces equally, and then at a suitable stage they are built into large stacks at sites easy of access. Each stack in the Fenland is reported as holding some ten thousand cesses, but local custom varies in this, as in the construction and naming of all the variety of the intermediate drying structures. Children and women often give a vast amount of part-time assistance in the very substantial labour associated with drying the turves, a process greatly dependent of course upon the weather and one that it is essential to complete well before the onset of frosts, for it is a major disadvantage of peat that the effect of freezing when moist is to cause rapid disintegration. A peat stack left exposed in winter is a sorry collapsed mess in the spring, and storage has to be with this in mind.

Despite this and its low calorific value, it had many advantages as a fuel, particularly cleanliness, attractive aroma, low ash-content, and the persistence of its slow combustion when left undisturbed on the hearth with easy renewal of the fire by the bellows at any time. As we have already indicated, in certain areas and circumstances it had the advantage of local availability. In the Fenland this was so for the medieval salt-extraction industry round the Wash, and somewhat surprisingly, at a much later date for the minor and local brick-works that occur here and there on the Fen margins. It is again valuable to have Will 'En's testimony that the brick-works at Ramsey Heights, Huntingdonshire, was fuelled by peat locally cut in the adjacent Lotting Fen. In the light of this it seems at least highly likely that such small brick-pits as those on Advenurers' Fen, Wicken and by North Breed, Wicken, were also supplied by locally cut peat (Plate *48*). This seems altogether more plausible than having to suppose the very expensive transport of wet clay on the one hand or of expensive hard coal on the other, for such small, out of the way, sites.

Over many of the Fens to which Skertchly refers as having peat diggings, drainage and cultivation have now removed all trace of it, but where, as in nature-reserves such as Chippenham, Wicken, Woodwalton and Holme Fen, there has been no recent destruction of the surface, there is clear evidence, particularly from the air, but also on the ground, that the whole extent has formerly been cut for peat, although of course the former trench and balks have weathered down to gentle ridges and hollows and the boat-dykes have filled-in and often almost disappeared. At my own earliest visit to Wicken I recall vividly that a large part of Adventurers' Fen was crossed by peat diggers' trenches that were all water-filled and still so steep-sided that jumping from one balk to the next across them was hazardous. Cutting here must have ceased with the 1914–18 war or thereabouts, but it persisted in neighbouring Fens for some years later, the turves being brought by boat for storage in a shed at the bottom of Lode Land, Wicken. It was about 1922 that I took the photograph of peat stacks on Swaffham Fen (Plate *29*), and Ennion's accounts of Burwell Fen will relate to about this time also.

Having regard to the presence of acidic bogs in the Holme, Woodwalton, Trundle Mere region and the known exploitation there of this preferred peat type, the conjecture is naturally raised of whether in the past, at least some surface development of acid bog peat might have been present also in the Reach, Swaffham, Wicken Fen area, but in the places where peat cutting has been minimal, borings never encounter any type of peat but black sedge peat with more or less brushwood throughout and one is forced to conclude that in large parts of the southern Fens it was this that was largely or wholly cut for fuel. Lacking all content of *Sphagnum*, it has never been exploited as horticultural peat.

13

The loss of the peat: shrinkage and wastage

As large-scale drainage of the Fenland became established in the mid seventeenth century under Cornelius Vermuyden, and subsequently gained in effectiveness, extent and speed, as sail pumps accelerated discharge and were themselves supplanted by steam engines and, in time, by diesel pumps, there followed one major consequence that was to prove crucial for the future of the black fens. This was the effect generally referred to at the time when the Fenland Research Committee came into being, as the 'shrinkage' of the peat. There was a general awareness that abstraction of water from the peat induced shrinkage, and there was much evidence before our eyes that the surface of the drained peat Fens was becoming lower; attention was called to this by Gordon Fowler in 1933 by a note on *Shrinkage of the Peat-covered Fenlands*. He cited historical evidence that the Drainage Boards had repeatedly had to lower the outfall drains from which water was raised into the rivers or major drains, and shewed how, in many places, the surface of the peat Fens was now actually below mean sea-level. His clinching argument was the vivid testimony provided by the 'Holme Fen post', an experiment initiated in 1848 by W. Wells, the enterprising landowner responsible for the drainage of Whittlesey Mere. He had set up in the Fen, about half a mile from the lake margin, a vertical measure consisting of a long cast-iron pillar resting on a solid iron cross-piece carried by oak piles driven into the underlying Jurassic Clay, and with the top precisely level with the fen surface of 1848 (Plate 35). Records were made at intervals of the extent to which the column came to project more and more above the descending peat-surface. By 1932 10.7 ft (3.5 m) had been exposed; although naturally the rate was greatest initially it has continued subsequently and the total thickness of peat has diminished overall from 24 to 10 ft (7.3–3.3 m). Magnitudes of lowering of this order are by no means rare in the Fens, if seldom so precisely recorded. When we were working in Wood Fen, near Ely about 1928 we had been struck by the extreme consistency with which the low uplands are ringed by a fen-drove that marks the upward limit to which fen drains extend, and the lowest limit of hedged fields. These droves must be very ancient for they carry the parish

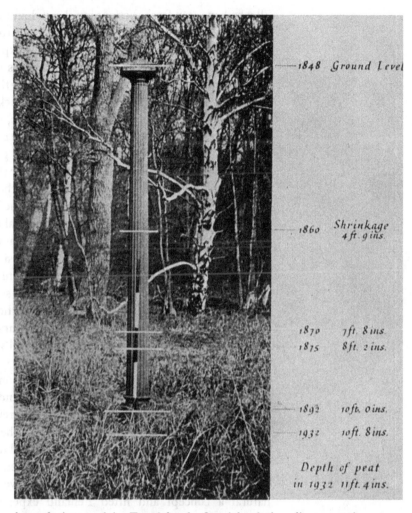

Plate *35*. The Holme Post as it appeared when photographed by G. Fowler in 1932. It shows the upper half of a long iron column inserted in 1851 on a firm base into the basal clay so that the top was level with the surface of the uncut raised bog at Holme Fen. In 1851 the rapid drainage of the adjacent Whittlesey Mere was begun and the marks indicate the progress of subsequent peat shrinkage and wastage recorded by emergence of the post. (See text and Fig. 36.)

boundaries, and in Fen island after island they lie upon the contour of + 12 ft (3.65 m) O.D. This must certainly have been the former upper limit of the peat-fen, whilst our own levelling at Wood Fen demonstrated that the surface in 1928 had then been reduced to Ordnance Datum level and locally even lower. It was from this Fen that Skertchly had reported that, shortly after its drainage was commenced, rates of 'shrinkage' of almost 2 in (5 cm) per annum had occurred.

Although it was natural to have pursued the use of this term 'shrinkage', it was essentially misleading, for besides the quick contraction in volume of the newly-drained peat that certainly was responsible for the early sharp fall in level seen at Wood Fen and Holme Fen, other processes were also operating and continued much longer. We have already ascribed peat preservation and accumulation to lack of oxygen below the water-table, and all who have seen fresh sections cut through untouched peat mires will have seen the astonishingly swift change in colour, from pale yellow or orange to dark brown or black, so soon as the peat is exposed to the air, a dramatic demonstration of its great oxygen demand. As water is withdrawn from a body of peat and air fills the spaces in it, there begin swift chemical oxidations followed by bacterial and fungal attack and breakdown by animal organisms. The peat being essentially organic, the ultimate product of all these processes must, to a very large extent, be carbon dioxide that diffuses into the atmosphere, becoming part of the great reservoir of that gas available for photosynthesis by vegetation. For this kind of peat loss, a total conversion to gas, the term 'wastage' is far better than 'shrinkage' and indeed the two carry different implications. The latter implies that however long water abstraction continues, all the original peat substance is still there and by inference one should be able to recover as complete a pollen-diagram from a drained as from an undrained peat deposit. This is, however, untrue, because the processes of peat wastage operate only layer by layer from above downwards, and as the water-level is lowered so all peat above it virtually disappears into thin air, so that the upper portions are absolutely lost and could *not* yield a pollen record. Likewise of course archaeological objects in the upper layers are *not* carried downwards still stratified in the compacting peat, but accumulate at the surface.

The importance of making this distinction between shrinkage and wastage came out early in the work of the Fenland Research Committee, not only in respect of pollen analysis but also with regard to the first hypothesis put forward initially to explain the shape of the silt roddons. This, based upon the concept of shrinkage through the full depth of peat below the silt flanges beneath the roddon, had to be abandoned in favour of an explanation accepting the principle of wastage from above. 'Wastage' was in fact so natural a concept and fitted Fenland experience so well that it was immediately accepted and is still in general use. It is a term indeed anciently employed in writing of Fenland drainage problems, and was quite properly used when recording the *disappearance* of peat in extensive areas following intensive and progressive drainage. It exactly corresponds to the experience of Fenland farmers who in their own life-times have seen what was deep black peat land altered as the plough progressively reached through it to the surface of the Fen Clay or, in more marginal areas, to the underlying Jurassic Clays or their mantle of Boulder Clay or gravel. The agricultural properties of the original black peat land and of the new mineral soils are so different, as also are the drainage problems they present, that they induce the adoption of new crops and new styles of management on the affected areas.

In addition to straight shrinkage or contraction and wastage, there remain as agents of peat loss the cutting of peat for fuel already described, that in some areas was responsible for removal of several feet of the uppermost layers, and the practice of 'paring and burning' as part of the process of reclamation for agriculture of freshly drained Fenland. This was a practice thought to have been introduced by Dutch settlers who came to the Fenland about the time when the two Bedford Levels and associated works were beginning to take effect. It consisted in shearing away the big tussocks of residual fen vegetation, gathering these *hassocks* together as they dried, into heaps which were ignited and left to burn on the peat surface. For the next century and a half this was the almost unvarying prelude to cultivation, beginning with cultivation of cole and proceeding to grain crops: there is even some evidence of its persistence in the accounts of the reclamations accompanying drainage of the Fenland meres in the mid nineteenth century. One may speculate that the *hassocks* might have been the great tussock-sedge (*Carex paniculata*) of more or less natural fen, the purple moor-grass (*Molinia caerulea*) from drying fen perhaps cut for hay, the tufted hair-grass (*Deschampsia caespitosa*) from fen margin grazings liable to flood, and most probably, in areas of acidic bog the tussock-building cotton-grass (*Eriophorum vaginatum*), whose fibrous remains are the *mabs* that so hinder peat-cutters everywhere (Plate *18*). All of these plants present severe problems to the improver and it remained common practice for their tussocks to be cut away by hand, so that even as late as 1868 there is record of the use of the breast-plough for this purpose on peat soils too soft to bear horses. Human labour was, however, supplemented or replaced to a considerable extent by the horse-drawn *hassock plough* and the *fen-paring plough*, special adaptations of the common plough that made a wide lateral cut to sever the roots of the tussocks. The burning of the dried vegetation of course provided an important mineral dressing for the new crops, but it carried the danger, severe in dry seasons, of starting peat fires that ate deeply below the surface, and were exceedingly hard to extinguish. As was increasingly recognised towards the end of the 1700s, the combined effect of the paring and burning was that 'it reduces the soil greatly, visible in the sinking of drained lands that have been pared'. Although the process is not recorded as other than one for the initial reclamation stage, it directly supplemented the processes of shrinkage and wastage that must have accompanied it.

The loss of 12 or 14 ft (3.6–4.3 m) of peat from vast areas of the black Fens has been of enormous and still-continuing significance for the whole economic life of the region, and has produced a variety of manifestations recognisable as one goes about the flat landscape whether on foot or by road and rail. Among the more commonplace is the way in which the buried bog oaks, in the language of the old fenmen, 'grow up through the peat', creating recurrent problems for the agriculturalist as they snag his plough or present him with fresh barriers to drainage as he deepens a field dyke and encounters a new massive oak trunk lying directly across it. The intensely heavy labour of removing successive levels of these tough old timbers has stamped its recollection in the minds of Fenland farmers although tractors have now replaced horses as the main agents of extraction. The farm workers engaged in such clearances noted with speculative interest the occasional recovery of stone or metal tools of prehistoric man, or of unfamiliar animal remains from the tree-level itself or the surface of the fen floor finally uncovered below it.

Some of the scanty trees that still grow in gardens or hedges on the black peat, and that have lived through a recent phase of rapid peat wastage bear odd testimony to it in the manner in which they now appear, perched upon some 4 or 5 ft of exposed root-system, with the base of the main trunk at that height in the air. The numerous stilt-like descending main roots of the alders allow them still to stay upright in this unnatural position. Likewise any old house, that was built for stability upon vertical piles, may now find itself, like the trees round it, elevated far above ground level, the front door high above its broken path and the steps that were initially added to maintain contact, now at last allowed to hang suspended on the building before the permanently closed door. All buildings seated directly upon the peat surface have necessarily subsided, more or less intact, as the land level has gone down, and not surprisingly, since the peat beneath them was far from uniform, with buried timber here and there, and liable moreover to unequal shrinkage, often in course of subsidence whole houses will have developed a drunken-looking tilt one way or another, with all the consequences of cracking and jammed doors and windows and the makeshift operations of supporting, strengthening and holding together (Plate *36*). The later generations of colonists of the peat-land, learning from this have built to improved designs, making use of basal supporting concrete rafts.

Plate *36*. The artificial course of the present River Great Ouse looking from near Littleport Bridge towards Brandon Creek. Since its construction, probably by the Romans, wastage of the surrounding peat has been so great that its surface is considerably below water-level in the river. (Photographed 1959.)

Plate *37*. The eastern bank of the River Great Ouse near Littleport. Wastage of the peat commonly has induced cracking and tilting of houses built upon it as can be seen by the angle of the roof of the nearest house with the horizontal bank of the River Little Ouse in the distance. The next house tilts towards the camera.

Older houses built upon peat where wastage has been great may demonstrate in an oddly effective way how great the total peat loss has been, especially those that stand beside some such means of comparison as one of the Bedford Levels, or the Prickwillow–Brandon Creek Ouse (Plate *37*), where the banks have necessarily had to be kept at their original height. Here one may see houses on the peat land just outside the bank with their roofs or bedrooms level with the boats passing along the water-way. It is easy to imagine in what danger the inhabitants of such houses stand at times of sudden and severe inundation, such as the flooding of 1947 or the constantly feared 'blowing' of the bank of the main drain: happily modern improvements in the overall Fen drainage have to some extent lessened the likelihood of such catastrophes. It is easy in the light of these effects to see why there are no villages built directly upon the peat: natural settlements always are marginal or sit upon the Fen 'islands' of gravel or clay. Where, as Astbury shews, a village such as Benwick is perched right on the boundary of the peat and the solid ground, it pays for this by cottages tending to lean backwards from the street into the fen behind. We have already drawn attention to the preferential siting of the Fen farms through the peat Fens upon the banks of roddon silts, and when we consider by contrast the great stretch of sands and silts making up the Marshland next the Wash, settlement and building are altogether more general. Against the total absence of churches on the peat, the silts carry an abundance of impressive and indeed well-known examples, among them notably Terrington St Clement's and Walpole St Peter. The late-Norman structure of Walsoken church gives, incidentally, a minimum date for the building of the great sea-wall whose presence is implicit in the names Walpole and Walsoken.

One of the more macabre consequences of the peat wastage is recorded in

Fenland Chronicle where Sybil Marshall's mother recalls that as children they used, on the way from school, to play in the churchyard of Ramsey St Mary's, where there had been no alternative to burial in the peat, and where wastage now exposed the coffins so that the children could reach down and touch them through the cracked soil. One of the less-believable stories about the wastage is that of Dr Lucas who writes of a Fenland five-barred gate, still suspended on its original gate-post, but now swinging so high above the road it once closed, that a pony and gig could drive below it! Some one had vainly spent a lot of labour and wasted a marvellously long gate post if this were true.

The wasting away of the mantle of peat from thousands of acres of the peat fen, and the contrasting immunity to compaction, wastage or burning of the mineral sediments such as silt and shell-marl, provided the geologist and archaeologist with a substantial bonus. As Gordon Fowler realised, it was this differential destructability that unmasked the silt roddons that meander now as raised banks through the regions of the peat fens. It was not that the roddon shape was in anyway altered by the process, but the wasting of the peat that formerly stood level with its crest and the terminal diminished stream along it, now revealed the original shape of the levées raised by tidal silt. Of course the same wastage gave opportunity to the cultivator to plough down the silt ridges so that in the low upstream sections the course of the roddon is apt still to be uncertain.

As the silt is largely immune from compaction and wastage, so are the molluscan shells and amorphous calcium carbonate of the shell-marl, so that where such deposits occupy channels once cut through the peat, they now stand proud of the surface of the wasted peat. Such are Fowler's 'old slades', like the straight Roman canal leading to the Fen margin at Reach, and others specifically recorded by Astbury near Red Mere and Streatham Mere. Where also the calcareous marl filled the bed of a Fenland mere, its drainage may now have left the white marl perched like a platform, a few feet above the present surface of the peat round about. This is clearly so at Red Mere as can be seen by anyone adjusted to the fen mode of seeing and giving attention to small differences of level that would pass unnoticed elsewhere.

The site of Whittlesey Mere shews up in the air photographs unmistakeably as a great area of shell-marl, but its present limits are not such as to disclose the same effect immediately. Nevertheless as one travels the road north east from the drained acid bog of Holme Fen on to the open cultivated area of the former mere, a rise of some feet becomes apparent, not, however, at the peat–marl contact but a mile or so towards the lake centre. There one finds quite an abrupt rise of about 4 to 5 ft (1.5 m), representing a boundary ridge, a step extending from north west to south east across the middle of the mere; all land to the north of this stands higher than that to the south though

Fig. 36. Transect from Holme Fen to Whittlesey Mere at three periods, 1400 B.C., A.D. 1850 and A.D. 1970, the last showing the consequences of differential compaction and wastage following the drainage of Whittlesey Mere in 1851, including protrusion of Holme post and development of a ridge across the mere bed above the buried junction of the raised bog and the edge of the Fen Clay hidden below the shell-marl. The 1400 B.C. reconstruction shows that the presence of raised bog limited the extent of the Fen Clay and points to deposition of thin fresh-water clays on the raised bog by flooding from upland where there was Bronze Age woodland clearance. These thin clay layers had become buried by 1850 in continued growth of raised bog that had meanwhile, in Iron Age or Roman time, set a margin to the calcareous mere at Whittlesey.

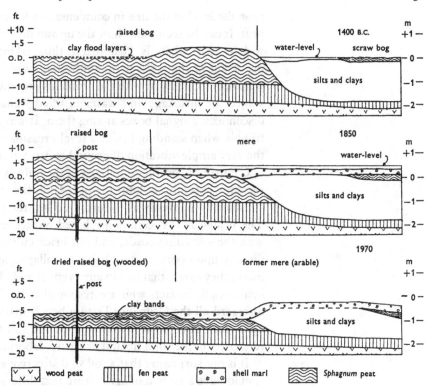

all alike is marl covered. The explanation of this odd geographical effect followed quickly upon systematic borings that shewed how, north of the ridge the marl rested upon the silts and clays of the earlier Fen Clay stage, and south of it wholly upon peat. Since the silts and clays have compacted so much less than the peat in the one hundred and twenty years since drainage of the mere, the contact between the two regions has come to be represented by a pronounced step. Peat shrinkage has not yet gone so far as to leave a similar step between the shell-marl and the peat at the lake-margin. We obtained evidence enough from the borings to warrant the schematic reconstruction set out in Fig. 36 to shew firstly the genesis of the mere between 1400 B.C. and A.D. 1850, and the consequences of drainage between that date and 1970. It is of course a reconstruction greatly helped by the measurements of Wells at the time of the drainage, and by the presence of the measuring post that he had the acumen and energy to establish for the benefit of his successors.

It is interesting that the railway embankment carrying the old Great Northern main line to Peterborough and York was initially built upon alternate layers of brush wood and turves allowed gradually to compact before addition of the ballast. All the same it was greatly affected by the drainage of Whittlesey Mere and vast amounts of material were added to raise the level of the line in compensation for subsidence of the underlying peat. It can be seen today how the up and down lines are spaced much more widely apart than is customary, thus taking account of the possible distortion of the track with continuing subsidence: the railway engineers were only too familiar with the problems presented by tracks built upon peat subject to shrinkage and wastage, tilted signals and telegraph posts and disconnected signal boxes among them. It was not hard to realise the basic trouble when standing beside a level-crossing on the peat-land and feeling the very ample vibration of ground as an express passed through.

Comparable problems to those of the railways naturally arose on the road system where, throughout the peat-land, the surface of every old road has subsided so unevenly as to forbid all but the most modest driving speed. The bridge heads, commonly based on deeper foundations, lose contact with the subsiding roads, and old brick culverts and bridges often based directly upon peat are very liable to collapse, and with them go the bench marks they carry, that had in any event already become sadly misleading. It behoves all, in fact, who are responsible for levelling in the Fenland to disregard all the Ordnance Datum bench-marks cut into structures based upon peat, and to carry instead their lines of levels out from the nearest reference on the solid upland or island, laborious as this may be.

It is not surprising that a soil so distinctive as that of the black peat fens should have a system of agriculture that differs substantially from those of either the silts of marshland or the various geological formations of the surrounding uplands. With adequate drainage they make extremely valuable farmland, though scarcely in the class of the marshland which commands prices as high as any in the country. The peat lands are almost exclusively arable, devoted principally to wheat, potatoes and sugar-beet for all of which, with generous added fertilisers, the yields are high. Grass leys are few, and what permanent pastures there are on the peat mostly lie in the 'washes', such as those between the Old and New Bedford Levels, where winter flooding is frequent. The interval between the two wars saw the introduction of what may be called 'market-garden crops', outstandingly celery and carrots, the cultivation of which is now a very significant part of the system: celery is a plant whose wild ancestors grow in marshy or even aquatic situations, and both crops are suited by the freedom from stones, friability and the ease with which adherent soil can be washed away before marketing. Of course the improved accessibility that has come with increased use of motor transport and the building of new concrete access roads by the County War Agricultural Committees of the 1939–45 period, played an important rôle in this development, and the quick cash returns are increasingly attractive.

Whereas initially the fears of the fen farmers were focussed principally upon the dangers of severe flooding, as the drainage improved and the peat has wasted, there arose anxiety that the land was over-drained for optimal summer growth and pleas grew for the retention of more of the winter water in the drains to act as reserve during a dry growing season. Local conditions favoured sometimes one and sometimes the other drainage policy and the opposing schools of thought came to be known respectively as 'dry-bobs' and 'wet-bobs'. It is a situation still unresolved.

As part of the price paid for improved drainage and more intensive cultivation there comes also increased susceptibility to another kind of 'fen-blow' than that of bursting dykes. It is the lifting of the dried particles of the friable peat soil by the wind, especially in dry periods during April and May after the spring sowing of crops. Strong winds will thus generate tremendous dust-storms of great extent that not only remove the top soil, fertilisers, seeds and seedlings and all, but deposit them in the freshly dug and cleared ditches, involving enormous labour in re-constituting the drains as well as in re-seeding the beet, onion, carrot or cereal crops. It is not unknown that two or three re-seedings are required. Here again there is no consensus as to systems of management (including maintained higher water-levels, 'claying', planting of shelter belts, and restoration of pasture) that could provide an economically practicable solution to the problem.

As the black peat progressively diminished, the lowered water-table over large areas came to lie thinly above the undulating surface of the impervious Fen Clay, whose reduced state was indicated by the name 'blue buttery clay'. Where there were basins in the peat surface so that the water could not drain away, anaerobic conditions likewise persisted inducing soil sterility and failure in the crop above: these had all to be brought into drainage connection with the dykes. This done, not only here but on the Boulder Clays beneath the more marginal land to which Fen Clay did not extend, farming has become essentially that of the upland farms. Gradually area by area, extending inland from the Marshland edge and seaward from the upland margin, as the peat wastes away, the farming régime comes to adopt the character of the adjacent peatless areas, after a short phase in which a mixture of peat and 'stronger' soil has given the best of both worlds.

Although the list we have given of interests affected, often seriously, by the peat wastage is already long, those whom it has concerned most continuously and expensively have been the many authorities, minor and major who, since the seventeenth century have been responsible for the maintenance of the Fenland drainage system and the passage across the Fen basin of all the river water discharged into it. Their complex and continuing difficulties, and the answers provided for them, seem, however, more conveniently dealt with when we have gathered a general appreciation of the whole, much-modified unnatural system of the Fenland waterways. This is described, though very briefly, in the next chapter.

14

Fenland drainage

The present appearance of the Fenland from the air is of a region threaded by a complex network of straight canals or drains, the major ones running directly seawards, but with many others entering them laterally and even establishing east–west linkages that cut across the general seawards slope. Large main rivers seen entering the Fens from the upland lose their identity on the plain and many alternative choices are proffered by the maps for the 'Old Nene' or 'former Ouse'. This situation holds not only for the major waterways but proves to extend downwards in scale to the minor rectilinear patterns of the drainage districts and field drains themselves.

It will be realised, therefore, how important it was to the field-work of the Fenland Research Committee that Gordon Fowler should already have made such headway in tracing the former natural waterways of at least the southern half of the Fenland. We have indicated the painstaking and effective procedure he followed, and how the identification of the roddon system enabled him to recognise the original main estuary of the southern Fen rivers extending seaward from Prickwillow via Outwell and Wisbech. In the course of defining the natural waterways, some few of which were still occupied by active streams, Fowler encountered others identifiable as man-made by their straightness, by their transection of hard mineral strata or by flowing counter to natural gradients. Such streams, as Skertchly pointed out, also often join a larger stream at an unnaturally large angle, often 90° or more, in a way no natural tributary ever does, and they often exhibit very unnatural 'river-capture' of existing streams.

Fowler's task has been most ably continued and extended by A. K. Astbury, so that now a very long register of artificial drainage works exists. Fowler had made it clear that the Romans had dug several of them, such as the Car Dyke, the Ely Ouse channel from Prickwillow to Brandon Creek and the massive straight roddon channel from Outwell to March. Astbury clearly regards more inferential evidence as suggesting that many more and a much greater total mileage were of the same Romano-British date.

Stratigraphic evidence of this period in a marl-filled slade or in roddon silt is, when found, clear enough, but there remain great numbers of clearly artificial channels for which no such dating is possible. Here the medieval historian or historical geographer comes into his own, a rôle filled in the Fenland Research Committee by H. C. (later Professor H. C.) Darby, whose monumental book, published in 1940, *The Draining of the Fens* is the effective compilation and presentation of all the historical evidence from medieval times onwards. With this available it would be presumptuous to give more than a brief sketch of the evolution of the Fenland drainage system, adding only sufficient to indicate the extensions of it since 1940.

There seems no evidence of Anglo-Saxon construction of waterways, and since they did not afford basis for taxable wealth, evidence of them is lacking in the Domesday Survey. Monastic houses were, however, to some extent involved with the creation of new waterways as is shewn by the thirteenth-century reference to 'Monks' Lode' leading to Sawtry Abbey, and there are other less well-known waterways of the same name, such as that flanking St Edmund's Fen at Wicken. It is not unlikely that many other of the ecclesiastic houses sited on the Fen margins, made at least local canals to link them with the main river-system, the end served, as with the Roman canals, being primarily that of transport. It should not be overlooked that such a use may well imply the need for staithes, locks and sluices: evidence of medieval or earlier structures of this kind might be worth the seeking.

There is certainly written evidence from the Middle Ages that the great monasteries made many agreements for the construction and maintenance of 'sewers' (i.e., drains), and Fen boundaries were marked by this and that 'new ditch' or 'new lode'. By this time the condition of such waterways was an important factor in sustaining the Fenland economy, but responsibility for it was obscurely shared by a great many individuals, townships and ecclesiastical authorities, so that neglect and disrepair were constantly alleged. Increasingly the crown sought to remedy this state of things by the setting up, from the thirteenth century onwards, of 'Commissioners of Sewers' with powers not only to report but to execute improvements and levy charges to pay for them. It was a system that permitted more than purely local control, and it developed into a somewhat formal and usual procedure. All the same, their activities were mostly concerned with repairs to banks and scouring of channels so that we learn of the construction of no considerable waterway intended to relieve flooding until Morton's Leam was cut between 1478 and 1490 to carry the waters of the Nene from their entry to the Fens just below Peterborough more directly to Guyhirne and so to Wisbech. Although the authority responsible for this project, Bishop Morton, was indeed a churchman, he was also a Privy Counsellor: his intervention seems to mark the beginning of effective national concern with the larger scale Fenland drainage. Although finally in 1600 there was passed *An Act for the recovering of many hundred thousand Acres of Marshes*, nothing

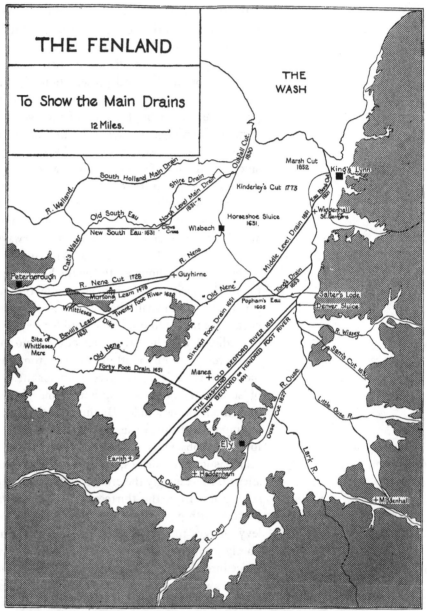

Fig. 37. Map given by H. C. Darby to show functional drainage of the Fenland in 1938: the approximate dates of the artificial cuts are given. Important subsequent additions have been the 'Relief Channel' between Denver and Lynn, and the 'Cut-off Channel' that intersects the Rivers Lark, Little Ouse and Wissey on the upland margin and conveys flood overload to Denver (see Chapter 14 and Fig. 15).

very satisfactory came of it. The first effective large-scale operation was that, which under sponsorship of the 4th and 5th Earls of Bedford, and the direction of the Dutch engineer Cornelius Vermuyden, created the so-called 'Old Bedford River' (1631), and 'New Bedford River' or 'Hundred Foot River' twenty years later. These were enormous straight drains, parallel to each other, capturing all the upland water of the River Great Ouse and carrying it directly north-eastwards from Earith, until they met 2 miles before rejoining that river at Denver Sluice. Thus the gradient was greatly steepened and the waters of the tributary Cam, Lark, Little Ouse and Wissey now only joined the discharge of the Bedford Ouse downstream of Denver. The final scheme had many features of importance and ingenuity. Sluices prevented ingress of salt water to the course of the Ouse above Denver and directed it instead up the New Bedford Level which now became tidal as far as Earith. Moreover, by providing that the outer bank of each Bedford River was higher by a few feet than the inner one it was possible in times of heavy flooding to open the sluices of the Old Bedford River at Earith and allow the discharge to fill the great area between the two rivers, some 1000 yards (900 m) across over most of its length of 18 miles (29 km). These Washes were thus at the same time a great reservoir of 2260 hectares for flood-water, a very important long extension of the discharge cross section of the rivers and an exceptionally fertile summer pasture. Although the Bedford rivers were the largest and most dramatic outcome of this period of drainage activity, many other considerable drains were cut at this time, the 'Forty foot Drain' that leads into the Old Bedford River via a sluice at Welche's Dam, 'Sam's Cut' that drains Methwold Fen into the Ouse, together with Bevill's Leam, and the 'Twenty foot River' taking water from the Fens east of Whittlesey Mere, the 'Sixteen foot Drain' and in the North Level of the Fens, the Peakirk Drain and the 'New South Eau'. The venture was a great success and the 'Gentlemen Adventurers' who had staked their capital upon it were rewarded by the agreed acreages of drained land, that still bear the name, here and there through the region, of 'Adventurers' Fen'. Cultivation of the drained Fens was widely undertaken, but progressively there now intervened the unforeseen consequences of shrinkage and wastage of the peat, nor have these ever subsequently been fully counteracted despite many new drains and extensive modifications and clearances through the last three hundred years. The basic problem is of course that the hard floor of the Fenland basin lies considerably below sea-level and the peat above it is progressively disappearing. Meanwhile, the silts and sands of the Marshland stand some 12 ft (4 m) or so above sea-level and, capped by their protective sea-banks, constitute a great embankment between the low peat Fens and the sea. Potentially all this diminished peat land is now a vast fresh-water reservoir. The natural and artificial rivers still pursue their original channels more or less directly across this basin to the sea, into which they still discharge gravitationally, in some cases at low tides only, and at some times assisted by pumping at the discharge sluices. One may readily imagine the first hazard that this situation presents, the danger of a 'blow out' of any bank that carries a tidal section of the drainage. The danger of this is naturally greatest when fresh-water floods coincide with high tides, especially if these are augmented by tidal surges in the North Sea

Plate *38*. A skeleton wind-pump from the Southern Fenland photographed by W. S. Farren. It raised water from the drain behind into that at higher level in the foreground, so keeping the land dry enough for the peat-cutting indicated in the background. An excavated 'bog oak' lies close to the pump. Such pumps were erected privately to drain quite small areas of fen.

Plate *39*. Drainage windmill that formerly stood on Harrison's Drove, Adventurers' Fen and that has now been re-erected on Wicken Sedge Fen, where from time to time it has been operated. Within the cylindrical case projecting on the near side runs a large scoop-wheel, the slats of which dip into the open water of a drain (here boxed over) and lifts it into a lead that discharges into the marginal fen-drain (marked by the reeds) that is behind the mill. (Photographed in 1964.)

Plate *40*. Drainage mill of a type similar to that on the Sedge Fen, that formerly stood on Spinney Bank on the north flank of the Fen, raising water from lower-lying agricultural land behind us into the drains that cross Wicken Fen and fall via Drainer's Dyke into Wicken Lode *en route* for the River Cam. When this picture was taken (about 1930) it had already been superseded by the diesel pump housed in the small shed to the right. The slats of the large scoop-wheel are clearly visible.

generated by prevalent strong north-westerly winds. There is still vivid recollection in the Fenland of the 'blowing up' of Denver Sluice in 1713 that deeply overflowed big areas of the region. Two centuries later we still had the condition that the whole South Level depended upon the security of the restored Denver Sluice, for this alone kept high tides *ordinarily* rising to +14 ft O.D. from entering the Ouse where the bank tops only reached +12 or +13 ft. No wonder there was anxiety when floods and *exceptional* tides coincided.

However, even more persistently present is the problem of rivers and drains that become steadily higher above the surface of the peat with which they once were level and that they drained naturally. By 1700 shrinkage and wastage of the peat were such that it was necessary to dig subsidiary drains converging to places on the banks of the main drains where windmills drove large scoop-wheels to raise the water over the containing bank. Thus in the early eighteenth century the Fen landscape was dotted with these decorative wind-pumps and they remained a common feature until the turn of the present century (Plates *38, 39* & *40*). We are fortunate to have the vivid personal record of one of the last men to have care of one. This was the invaluable Mr Edwards whose interest in and devotion to his wind-pump come vividly through the pages of *Fenland Chronicle*. Decorative they might be, but theirs was a restricted efficiency: even with enlarged scoop-wheels

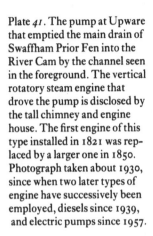

Plate *41*. The pump at Upware that emptied the main drain of Swaffham Prior Fen into the River Cam by the channel seen in the foreground. The vertical rotatory steam engine that drove the pump is disclosed by the tall chimney and engine house. The first engine of this type installed in 1821 was replaced by a larger one in 1850. Photograph taken about 1930, since when two later types of engine have successively been employed, diesels since 1939, and electric pumps since 1957.

their lift was at most a few feet and they were at the mercy of gales, calms and prolonged frost so that they had often to be idle when most needed. The introduction of the steam-engine greatly reduced all these limitations and the introduction of the centrifugal pump about 1850 accelerated their adoption. A few examples happily remain of the beautiful Victorian vertical beam-engines, each within its narrow upright brick engine-house that with its accompanying tall chimney came to be a Fenland feature almost as distinctive as the wind-pump (Plate *41*). These in turn have given place to diesel engines, their compact shapes in lower square buildings often alongside an older engine-house. They too register change, for once supplied with diesel oil by water, they now tend to be serviced by lorries coming over the improved fen roads, and increasingly from 1948 electric motors under automatic operation and driving mixed flow or axial flow pumps have been adopted. Technology here is well ahead.

Mr Doran, for so long Chief Engineer of the Great Ouse Drainage Board, has effectively summarised the chronology of these changes:

(1) Windmill and scoop wheel⎱
(2) Windmills in series   ⎰   Until 1820
(3) Steam engines and scoop wheels   1820–51
(4) Steam engines and centrifugal pumps   1851–1947
(5) Diesel engines and centrifugal pumps   1913 onwards
(6) Electrical motors and mixed flow or axial
    flow pumps automatically operated   1948 onwards

We have to envisage this picture of improved efficiency and change as accompanying a linked development of a new subsidiary low-level system of drains traversing the well-cultivated fields and collecting the drainage water at the pumping stations. Thus we now commonly find, outside the bank of the main drain a corresponding 'Commissioners' Drain' at field level and find one or more substantial 'Engine Drains' converging directly upon the pumping station.

The more efficient drainage has become, the further of course has the peat-level subsided, not only with the consequences already described, but with the severest difficulty, and indeed jeopardy, put upon maintenance of what I always visualise as the 'perched' system of water-ways, crossing the peat fens like an elevated aqueduct, which is indeed what they are. As the river banks have risen higher above the wasting peat land, the load on their foundations has steadily increased, and as they are only of peat, now itself drying out and in any case compressible, it is not surprising that the average rates of subsidence of the bank-tops of the main drains should be reckoned by the drainage engineers to be as great on average as 0.75 in (1.9 cm) per annum. Although the upper banks themselves made of peat, have often been breasted or capped with Gault Clay, the newly-exposed bases are sealed internally by no more than the silty muds of the channel itself, and the drying peat behind is increasingly liable to crack and leak under the increased head of water. This no less than the overflowing of the banks offers the danger of a catastrophic burst. To this one adds the problem that increasingly the load of the embankment becomes less evenly distributed, so that the banks tend to tilt outwards and become unstable. Nothing remains but to attempt to raise and waterproof the banks, but this is difficult, expensive and continuing. The raised banks require broader foundations with great quantities of fresh material, but more serious than this is the fact that in many instances the bank was built too close to the channel to allow any expansion of the base on the *inside*, whilst on the *outside* the low-level drains, and commonly important roads and attendant buildings lie directly in the shelter of the bank. To undertake their removal to give room for the widened bank is extremely expensive, and equally daunting is the prospect of having to re-site all the sluices and pumping stations along the bank. Nor is this all: the heightened bank must be itself sealed with imported clay, failing which a deep continuous trench must be dug down from the bank face and be filled with puddled clay. With such conditions maintenance must have something of a hand-to-mouth quality despite all the experience of the responsible engineers: it is only surprising that recent breaches of the banks should have been relatively infrequent and of limited severity.

Now that the essence of the problem has been so unquestionably demonstrated as that of a 'perched' drainage system crossing the vast

shallow basin 10 or 15 ft below it, the basic remedy has suggested itself of preventing the rivers of the uplands from discharging themselves primarily or at all through the elevated waterways that cross the South Level. The massive upland floods of 1936, 1937 and 1939, exceeded by that of 1947 and that of January 1953 notably augmented by the North Sea tidal surge, gave urgency and a scale of dimensions to the planners. The first element of the remedy has been the construction of a 44 km Cut-Off Channel that passes along the Fen margin from Barton Mills, near Mildenhall, intercepts the rivers Lark, Little Ouse and Wissey and that can entirely bypass their floodwaters round the Fen basin to re-enter the tidal Ouse below Denver.

Severe as are the twin problems presented by the floods discharged into the Fens from large upland catchments, and their safe carriage across the deepening concavity of the peat fens, they both are ultimately subordinate to

Plate *42*. Oblique air view looking north east across the new and old sluices at Denver, Norfolk. The 'Cut-off' Channel enters from the bottom left: it is controlled by an impounding sluice and communicates when necessary by the two control sluices (on curved cross channels), with the right-hand terminal extension of the River Great Ouse. This extension is separated by a massive head-sluice from the 'Relief' Channel into which the Cut-off discharges and which ultimately reaches the River Ouse near Lynn. The left-hand terminal of the River Great Ouse is crossed by Denver Sluice itself that diverts all tidal water coming up the Ouse from the sea into the New Bedford Level. At the top of the picture is the village of Salter's Lode, where the attenuated *Old* Bedford River also enters the tidal Great Ouse. (Photograph by Department of Aerial Photography, University of Cambridge.)

a third factor, that of their final disposition in the strongly tidal waters of the Wash. Erosion of soft sedimentary rocks along the eastern coast of England and the prevalent tidal movements constantly bring a burden of fine mineral matter into the Wash where it is deposited at slack tide, filling the estuary with shoals and banks, especially in the eastern corner where man has now introduced the enlarged outfall of the Ouse. With the flood tide fine silt in suspension is carried upstream, where a proportion of it remains since the ebb tide's lessened velocity will not support it. The silted-up outfall channels with their diminished cross-section consequently have constituted over three centuries or more a further hindrance to efficient Fen drainage, and all large-scale plans must take account of them. It has been of assistance in preparing the latest of these to take advantage of results obtained by the use of a large-scale tidal model of the Wash operated in Cambridge by the River Great Ouse Drainage Board, particularly between 1935 and 1940. In this not only was the morphology of the Wash reproduced at any given time, but the rivers entering with appropriate fresh-water flow, and in the estuary itself sea-water and the actual silt of the Wash river outflows. By a suitable system of gearing large floats were operated in a reservoir tank so that a twelve-hour tidal cycle could be replicated in a few minutes. River rise and fall could be recorded on the model and compared with those obtained directly in the field at corresponding times and sites on the rivers at different distances from the outfall. Furthermore, it was now possible to shew how the present disposition of tidal silting had come about, and to test the consequences of different proposed engineering treatments of training walls of varying length and separation, of new river cuts and so forth.

Substantially as a result of experience gathered in this way work was begun in 1954 on construction of a *Relief Channel* 11 miles (18 km) long, lying alongside the existing Lynn Ouse downstream from Denver and re-entering it through an enormous tail-sluice with seven openings, each 20 ft (6.5 m) wide, controlled by steel gates that weigh in all, over 400 tons. At times of flood discharge the tail sluice opens when tide level falls sufficiently, closing as the making tide comes up but meanwhile having emptied this very large reservoir of its accumulated content. The top water-level in the relief channel (+6.5 ft, +2 m O.D.) is such that flood discharge from the Ely Ouse and the *Cut-Off Channel* is thus not interrupted by the tidal cycle. Furthermore, there is the great advantage that the bed of the *Relief Channel* is not cut in soil susceptible to shrinkage or wastage, and existing gradients in the Ely Ouse can be steepened to take advantage of the improved discharge. The *Relief Channel* came into operation in 1959, the *Cut-Off* in 1964. Together these great engineering projects go far to support the claim that the Water Authority 'can now exercise complete control over flood water passing through the South

Level'. They are necessarily associated with substantial continuing work in keeping open the primary shipping and outfall channels through the Wash, a consideration that brings a timely reminder of the close and continuing domination of the Fenland's history by the sea.

We have seen how Fenland was indeed created by the general rise of ocean-level that followed the last Ice Age, and how it has twice thereafter suffered from considerable marine invasions, the first resulting in deposition of the Fen Clay, possibly to be regarded as the culmination of the initial waterlogging, and the second the double marine incursions of the Iron Age/Romano-British period, that deposited the great 'Marshland' belt of silts and clays and created the silty river levées that ultimately became roddons. There is some evidence suggesting renewed, though minor, rise in sea-level during the Middle Ages and it is perhaps foolhardy to assume that we are now to enjoy continuing total stability of relative land- to sea-level. Not only of course may the height of the oceans vary overall as evaporation, ice-melting and changing ocean floor dictate, but the earth's crust twists and rises and falls in response to altering loads upon it. Thus raised beaches round the Scottish coasts bear witness that since the ice-burden melted away, northern Britain has risen considerably above its former height. One naturally wonders whether south of the hinge-line that crosses this country from about the Humber to North Wales, a corresponding depression of the earth's crust may have happened, and worse still, might still be going on. There have been indications, such as those provided by continuous tide-gauge readings at Lynn and Felixstowe, or by Ordnance Survey triangulation, that this might exist but the evidence is slight and confused by the added complexities of greater frequency of tidal surges. The building of the Thames Estuary barrage emphasises how close to the danger-line are the cities built upon the great coastal ports of the southern North Sea, and the Fenland's past history warns that it would be foolish to ignore the sea-level component in any calculations involved in safeguarding the future of this coast.

15

Ancient crops, natural and cultivated

The Fenland now exhibits a modern highly-specialised agriculture appropriate to some of the most highly valued farmland in Britain, but since this account chiefly concerns the lost or vanishing Fenland, it is limited to crops now entirely lost or out of favour and unfamiliar.

A great many of the products of the undrained or half-drained Fens require no labour other than that of their collection and transportation, and it has been convenient to treat these separately. Their diminished significance has followed naturally upon the progressive reduction of undrained Fenland. It is otherwise with the drained and farmed Fenland: here crops have primarily been lost through the operation of economic factors, lessened demand for the product, increased labour and transportation costs and outside competition. It seems worthwhile, whilst the evidence and recollection still persist, to assemble some of the main features of these past or passing elements in Fenland economy.

The sword-sedge (*Cladium mariscus*), one of the ancient natural crop plants of the peat-fens, is quite uncultivated, exactly like the great New

Plate *43*. The giant sword-sedge, *Cladium mariscus*, in fruit in a ditch beside the Main Drove, Wicken Sedge Fen, 1956.

Plate *44*. The sedge crop being brought home by barge along Wicken Lode, by the Fenkeeper William Barnes and his son, Henry, who was to succeed him. Note the clearness and rich aquatic flora of the Lode, and on the far side of the Lode, Wicken Poor's Piece, at that time still so regularly cut for sedge that it shews no bush invasion.

Zealand flax, that is still cut as a wild crop from the lowland swamps of that country. *Cladium mariscus* is, as we have said, a natural component of the vegetational succession in open waters, preferring a depth of a few inches of water, and it there produces its linear leaves to a length of as much as 9 or 10 ft (3 m): these grow basally and bend over at a height of 3 or 4 ft (just over 1 m) making a very close stand, and they are ferociously armed along the margins and the keel with sharp teeth, so that if you should grab them to keep a foothold they will cut your hand to the bone! A fall and a soaking are much preferable. These leaves are produced from a substantial system of vertical stocks and lateral rhizomes that have fibrous roots in the aerated top layers and fat spongy black roots that penetrate into the anaerobic muds or peats below, themselves supplied with the oxygen they need by diffusion through the profuse air spaces extending from the bases of dead leaves above water-level. It is waterlogging of these broken leaf-bases that stops the plant from invading deeper water. The roots and rhizomes are largely responsible for the quick accumulation of peat as can be seen by their abundance through great depths of fen peat, as at Wicken Fen. At this famous site the sedge-dominated communities have persisted well into the stage where peat has grown above water-level, though its growth is better in the old dyke channels. The leaves are evergreen and persist individually for several years and they have traditionally supplied a thatching material of durability far exceeding that of cereal straw and even of the 'Norfolk' reed (Plates *44*, *45*, *46* & *50*). The sedge crop is cut by hand- or mechanical-scythe in late summer and autumn at intervals of three or four years: more frequent

Plate *45*. The former use of the sedge crop is illustrated by this old picture of one of the cottages in Lode Lane, Wicken thatched with *Cladium*: spare sedge is piled behind the garden fence. The cottage has now gone and thatching with sedge has ceased.

Plate *46*. Sedge thatch at Swaffham Prior, 1974: the close view shews the rough texture given by the harsh durable leaves of the *Cladium*. (Photograph by W. H. Palmer.)

cropping is damaging and in the past has led to conversion of 'sedge-fields' into communities containing much purple moor-grass (*Molinia caerulea*), dwarfer sedges, rushes and flowering plants that cannot exist under the uncut sedge with its thick mattress of dead, but persistent, old leaves propped up by the living shoots (Fig. 38). The former value of the sedge crop is recognisable in the fact that conditions governing the use of the 'Poor's Pieces' include limitations upon the commencing date of sedge-cutting and embargo on the employment of assistants. Wicken village used formerly to exhibit houses thatched with sedge, or with sedge-capped reed, but the introduction of corrugated iron has reduced demand to small proportions, although the management committee for the Fen still maintains some sedge-fields. For some years recently the sedge had a market on the Newmarket training stables where the chopped, sharp-edged leaves spread on the ground gave foothold to the horses in spells of prolonged frost. Not only was the thatching material carried by boat to neighbouring townships, but in times before the prevalence of daily papers, the dead *Cladium* leaves were imported as kindling material.

The plant appears to flower only in wetter places and in warm seasons, but it makes handsome conspicuous inflorescences and sets viable seed. There is certainly every reason to suppose it to be a native plant present at Wicken for the last five or six thousand years (Plates *43* & *47*).

Fig. 38. Diagrammatic bisect through the 'mixed-sedge' community at Wicken Fen, the source of the main sedge crop, and the consequence of cutting at about three-yearly intervals. The general vegetation level is about 1 m and is determined by the bent-over leaves of the sedge *Cladium mariscus* (C), through which project occasional shoots of reed, *Phragmites communis* (P). Two small tussocks of the grass *Molinia caerulea* (M) are shown and from one grows a plant of hemp agrimony, *Eupatorium cannabinum* (Eu). A few plants of the marsh pennywort are shown, *Hydrocotyle vulgaris* (H). The mattress of dead sedge leaves that darkens the ground surface has been omitted.

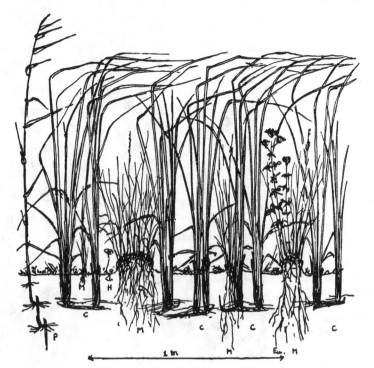

Plate *47*. A sedge-field in the main sedge-fen at Wicken, shortly after the disastrous fire of 1929 had displayed several important ecological and historic features. The white shoots are the killed and bleached leaves of the *Cladium* being pushed upwards by growth from below. Note the absence of sedge-shoots below the charred bush of guelder-rose: under the canopy of the bush the *Cladium* had been killed out and the absence of a mattress of dead sedge below it, to act as kindling, largely saved the guelder-rose. Note the parallel lines that indicate the widespread former peat cutting from which state the sedge-field established itself.

Phragmites communis, the 'common reed' or 'giant reed' and the largest of British grasses, has in all likelihood been a good deal more widespread source of thatch throughout the Fenland as a whole than the sword-sedge. Equally a native plant it occupies much deeper water than the sedge, rooted in the muds even 3 ft (1 m) or more below the water-surface. It forms very often an outer zone to the reed-swamp, sheltering below it totally submerged and floating aquatic plants such as the pondweeds, duckweeds and water-lilies, but it grows 3 m or more tall and has an astonishingly dominant life form. An apparently identical reed-swamp occurs over a great range of latitude and longitude, so that for instance it covers vast areas of the Danube delta where it exhibits the interesting behaviour of becoming detached by high floods as large foating islands, the *Plav*, that gently float into sheltered locations and then continue independent growth, possibly even supporting some small human occupation. Unlike the sword-sedge, the aerial part is a cylindrical tube of concentric leaf-sheaths around the flowering axis. Anthony Day's evocative fen pictures, no less than Chinese waterside paintings on silk, owe much to the shiny inner surfaces of the leaf-sheaths of the reed. These are so polished and so long that in a good wind they allow the big leaf-blades to twist round and fly downwind like pennants, an effect that generations of artists have made use of. A twist of the fingers brings the leaves back at once to their original spiral disposition. The whole aerial shoot is an annual growth so that if left uncut the haulms of the current year are admixed with the older, partly decayed leafless shoots of

Plate *48*. Former brick-pit on the eastern margin of Wicken Sedge Fen to shew the fringe of the giant reed, *Phragmites communis*. The very sudden deepening of the sides limits invasion of open water by the reed: although occasional shoots trail outwards they will not establish themselves. White water-lilies can, however, root at the greater depth and establish themselves along with milfoil, bladder-wort and other submerged aquatics. (Photograph by W. H. Palmer.)

earlier seasons. The best quality reed crop is accordingly cropped annually, being cut close above water-level and tied into sheaves. It will be appreciated that harvesting and maintenance of the reed-swamp are easier where there are considerable stretches of open water and this explains why, having just such conditions in the Broads, Norfolk became recently so much associated in the public mind with the crop that it has acquired the name 'Norfolk reed'. However, we may be sure that the plant has had a similar long history of economic use throughout the Fenland wherever good stands of it grew naturally, as for example in the undrained meres and less-used drainage ditches. At Wicken, reed was cut for thatch mainly from ditches and lodes, but the excavation of the large shallow artificial mere in Adventurers' Fen and deliberate waterlogging of former areas dug for peat have provided extensive reed-beds that are fairly regularly cut and provide a welcome augmentation of funds for upkeep of the fen as a whole (Plate *48*). These reed-beds of course also derive great importance in this instance from the fact that they are the characteristic habitat of numerous birds typical of the fens, such especially as the reed-warbler.

There really is no evidence that the dominant plant of the outermost reed-swamp, the true bulrush (*Scirpus lacustris*) ever formed more than a local crop, but its employment from at least medieval times for making rush-mats, palliasses and the seating of country chairs, strongly suggests a continuing local usage wherever it forms good growths, as it does in clear lodes and streams towards the Fen margins. It is a very distinctive plant

with its transparent ribbon-like leaves floating deeply below the water-surface: the aerial part, a dark-green tapering cylinder, is really the flowering
haulm that carries a cluster of brownish flowers near the pointed stem apex,
but bears no other leaves (Plate 4). It is confusing that the name 'bulrush'
should so often be used for the 'reed-maces' or 'cat-tails', whose velvety
fruiting heads were so often used to ornament late Victorian drawing rooms
and halls.

An extremely unusual crop, and one that may never again be called for,
was taken from Wicken Fen in the first years of the 1939–45 war. It had been
dismayingly discovered that our supplies of the evenly-burning charcoal
used in certain shell-fuses, derived entirely from sources now in enemy
hands. They came from the wood of the alder-buckthorn *Frangula alnus*, a
bush shewn by botanical survey to grow abundantly in very few British
localities. Among these, however, was Wicken Fen, where, as it chanced,
widespread colonisation by this bush was threatening to destroy many of the
Fen's most typical and scientifically important communities. Very large
quantities of alder-buckthorn wood, cut from the many stemmed stools at
ground level were accordingly taken from the Fen during the early war
years, but it is not a product for which demand has continued and bush
clearance, largely concerned with this species, has had to be undertaken in
the post-war years without financial help from sale of the all too abundant
wood. The related purging buckthorn (*Rhamnus cathartica*), that also
grows, less freely, on the Fen was never used for charcoal, no doubt basically

Plate *49*. The crop of giant
reed (*Phragmites communis*) in
recent years has become a
valuable source of income to
the National Trust's committee of management. The
stacked bundles shewn here
beside the head of Wicken
Lode are of high quality, much
esteemed for thatching and are
largely provided from the
dykes and from wet parts of
Adventurers' Fen. (Photo-
graph by W. H. Palmer.)

Plate 50. A fine example of modern thatching at Wicken. The smooth thatch of the reed (*Phragmites*) is capped by a ridge of the harder-wearing sedge (*Cladium*), both traditional and local fen crops. (Photograph by W. H. Palmer.)

because its different wood anatomy affected the character of its combustion.

Aside from this singular episode the trees and bushes of the natural fens have been subject to little or no economic exploitation. The *Salix* that grows so freely in the Fen carr is always a sallow, *S. cinerea* and the osiers seem never to appear naturally there. These are species of willow, *S. viminalis* (common osier), *S. purpurea* (purple osier) and *S. triandra* (almond osier), whose straight, pliant and clean shoots, induced by repeated pollarding and cutting, supply the material of a very widespread, if diffuse, industry. This produced for agriculture, horticulture and domestic use a great range of hampers, baskets, skips, tough screens and a variety of durable and light containers. Osiers also continue in very general use throughout the Fenland as supplying mattresses and stakes for securing the safety and repair of river banks and drains, notably those exposed to tidal scour. The osiers in general seem to favour more mineral soils than pure fen peat, but they were grown abundantly by the Fenland rivers until plastic and paper technology superseded them, as iron and asbestos roofing had superseded thatch.

Mention of the osiers reminds us of the particular and important rôle played by the Fenland in the economy of the many settlements close to its margin. The theme is admirably developed by J. R. Ravensdale for the villages of Landbeach, Waterbeach and Cottenham, pursuing the evidence right through from late Roman time to the present. He shews how a substantial arable agriculture on the higher mineral ground was accompanied by a primarily pastoral economy in the Fen-side marshes and meadows, with a generous supplementation by such typical Fenland products as turf, thatch, osiers, wild-fowl and fish. To some degree this ambivalence of resources was an insurance against the opposing threats of drought and flood, the same kind of alternative resource that coastal fisher-farmer people had taken advantage of from Mesolithic times onward. The value always set upon the summer usage of the Fens can be judged from the way in which each Fen-edge parish extends as a strip down into the Fen, and by the sharp disputes of the ecclesiastic houses on their rights to the Fen meadows and pasturage. To judge by the few remaining examples, these must have been direct derivatives of the natural Fenland communities modified by their pattern of usage but still essentially composed of their original plant species of grasses, sedges, rushes and a range of herbs tolerant of some cutting and grazing.

The case is quite other when we turn to consider the impact upon great areas of the central Fenland of the successful large-scale drainage of the seventeenth century that now for the first time permitted wholesale arable cultivation of both peat and silt land. Several good authorities shew what an important supplementary rôle in this process was played by one particular crop-plant, the rape, 'cole-seed', or 'rape-seed' that had a very high reputation as introductory crop on the freshly available land. It is a genetic form of cole *Brassica napus* var. *arvensis*, and since the wild cabbages are sub-maritime plants tolerant of salt, it is not surprising that rape should have had particular success upon reclaimed salt-marshes – indeed it is used still by the Dutch as first crop upon their own reclaimed polders. This plant was grown from classical times for the oil-content of its shiny blue-black

Plate *51*. Young carr of alder-buckthorn (*Frangula alnus*) that has invaded and killed out the community of giant sword-sedge (*Cladium mariscus*), the persisting dead leaves of which still hang in the crotches of the dense buckthorn, a witness to the reality of the vegetational succession. (Photograph by G. E. Briggs, 1928.)

seeds, and was the main source of vegetable oil throughout those parts of Europe outside the south-east and Mediterranean regions where sesame and olive grew. Its introduction to the Fens has been conjecturally associated with a settlement near Thorney of French Huguenot refugees who came here via Holland. It was written of them by A. Young (1805) that they 'began to sow a kind of wild cabbage called colza, from which they extracted an oil, which is not only very useful for lamps, but also is used in the preparation of wool'. At Thorney they were presumably growing rape upon reclaimed peat fen for it is to them that is attributed the fen-paring plough used to facilitate the destruction of the top sods of fen vegetation as a first step in introducing the crop.

As in mainland Europe, so in the Fenland, colza-oil production became extremely profitable and widespread during the middle of the nineteenth century, by which time the crop was 'largely grown both for sheep feed and for crushing purposes, a 1000 tons of oil being said to have been shipped yearly from Wisbech, no less than seven mills being employed crushing it from the seed'. It seems probable that the plant was winter-sown and, after its rather sudden ripening, was harvested and threshed in late summer: the residue from pressing provided a valuable cattle food, anticipating the use of cotton-seed cake. The browish-yellow oil was burned in lamps with a broad wick. Its use for illumination had decayed, however, throughout mainland Europe as well as Britain by the end of the last century, displaced in turn by the use of paraffin oil, coal-gas and finally electricity: the percentage of land devoted to its cultivation as an oil crop correspondingly declined, but its use for both oil and fodder has very recently increased again, though separate varieties are grown for the two purposes.

The prevalent public association of the hemp, *Cannabis sativa* with its employment as a drug, and with legislation to control this, have led to some failure to appreciate the ancient more respectable use of the plant as a fibre crop, one grown in considerable amounts throughout the southern half of the British Isles, including the Fens. The hemp is a plant of western Asia and India grown from *c.* 1000 B.C. for its bast fibres and oily seeds, although Scythian tombs from the Altai in the Russian steppe contain apparatus suggesting that inhalation of the smoke from burning hemp seed was practised in the steppes during that millennium. The plant was grown in classical times throughout Greece, Asia Minor, and the eastern Mediterranean, its fibres being used for cordage and textiles. It seems probable it was to the Teutonic people of eastern and central Europe that we owe introduction of the crop to Britain, particularly when the waves of folk migration brought to this country the Anglo-Saxons, with their entrenched agricultural traditions and their use of a large ox-team plough that was capable of tilling our stiffer clay soils. Hemp now became an extremely widespread crop, figuring in church assessments along with flax in the fourteenth century and appearing abundantly in place-names such as 'Heneplond' or 'Hemplond'. Hemp was now the chief textile fibre for the peasantry through the Middle Ages, and 'homespun' is only a modified 'hempenspun'. It is an astonishingly vigorous plant and, although only an annual, during the course of one short growing season will achieve twice the height of a man. It requires a temperate climate whilst growing, and deep, fertile and moist soils, whilst the retting of the stems prior to scutching, as with flax needs a good water-supply. These needs were largely met in the Fenland. Even before the Dutch drainage, a commission on tithe collection in 1591 at Whittlesea had ruled 'the grower to pull and use the hemp tythe as their own (except the watering thereof) and to carry the tenth sheaf unto the Parsonage gate . . . no tythe for hemp seed', and Drayton's *Polyolbion* of 1613 had referred to 'Hemp-bearing Hollands Fen', i.e., the Fenland of South Lincolnshire, afterwards reputed to grow 'the best Hempe of England'.

In 1685 the early benefits of the Bedford Level were celebrated in elaborate verse

> Here thrives the lusty Hemp, of strength untamed,
> Whereof vast Sails and mighty Cables fram'd,
> Serve for our Royal Fleets.

This conveys a timely reminder of the importance of hemp for these purposes, so long as the Navy depended on sail. Indeed legislation had been in force since the reign of Henry VIII to the effect that every person occupying land for tillage should sow yearly a quarter-acre of flax or hemp seed for every 60 acres (10 hectares) of arable that he possessed. In fact during the sixteenth century almost 15 per cent of the sown acreage of the Norfolk Fens was under hemp, and right through into the nineteenth century all parts of the Fenland, both silt and peat bore crops of the plant. It is a far cry from this to the surreptitious and illegal harbouring of a few plants of 'pot' in some back-garden, although the Fen folk were doubtless aware of some of the effects of hemp as a drug, since it appears that sometimes hemp seeds were smoked with tobacco and also that workers occupied for long hours together in the hemp crop became exceedingly drowsy.

Hemp was cultivated throughout the world during the nineteenth century, suitable deep soils and a temperate to warm growing season allowing production of giant plants. By 1900 London was the centre of the world's market for hemp, but subsequent competition from the products of other fibre plants, for example jute and manilla, and from synthetic fibres have very greatly reduced its importance. Although the hemp seed extraction yields an oil used for making varnish and soap, there seems no evidence that the plant was grown in the Fens except for fibre. However, when during the Second World War, our overseas supplies of palm oil and whale oil were threatened, we turned again to the possibility of raising a vegetable oil crop at home. Hemp was one of the crops considered and it was interesting to find that the plots of Fen soil chosen for trials turned out in fact to be, within the recollection of the oldest locals, those also formerly favoured for hemp.

I have not thought it necessary to include any account here of the Fenland cultivation of the flax, *Linum usitatissimum*, since it remains a fairly familiar crop elsewhere in Britain. It should, however, be noted that hemp and flax appear always to have been strongly associated, since requirements of soil, cultivation and retting are similar for both. '*Canab et lin*' were a single item for fourteenth-century assessment of church dues, and in the eighteenth century at Spalding on April 27 there was annually a special fair devoted entirely to these two fibres. Right through to the agricultural surveys of around 1800 the two crops were grown in the same areas of the Fens and in similar amounts.

A plant now far more seldom seen in the Fenland than formerly is the opium poppy, *Papaver somniferum*, a striking and easily recognisable robust plant with glaucous waxy leaves and stems, large white, red or purple flowers and finally the large spherical capsules that, like the rest of the plant, yield a white milky latex wherever it is cut or broken. It is a species now not known as a wild plant although the closely related *Papaver setigerum* is found around the Mediterranean. The long history of the poppy as providing an opiate and sedative seems confirmed by identification of its seeds in large quantity, along with those of other medicinal plants, in excavations of the Neolithic lake-dwellings. There is good evidence of its use in the Levant in the Late Bronze Age, the Minoan civilisation employed it extensively, apparently in religious ceremonial, it was written of by Homer long before the birth of Christ, and Greek familiarity with it was such that they depicted its globular capsules as symbol in the hand of Hypnos, the god of sleep.

It was introduced from the near and middle East to India and thence, quite late in its history, to China where opium smoking became appallingly prevalent. This widespread abuse was based on importation of the drug from India and it was to support this trade that Britain fought the shameful 'Opium wars' of 1843, 1856 and 1860. The developing capsules contain a great wealth of different alkaloids, Morphine, Codein and Laudanum among them and the opium poppy crop remains the primary source of these extremely valuable drugs. Before the days of advanced pharmaceutical separation and purification of the alkaloids, preparations of the poppy fruits were made as pills, powders, extracts, tinctures, syrup, salves and plasters,

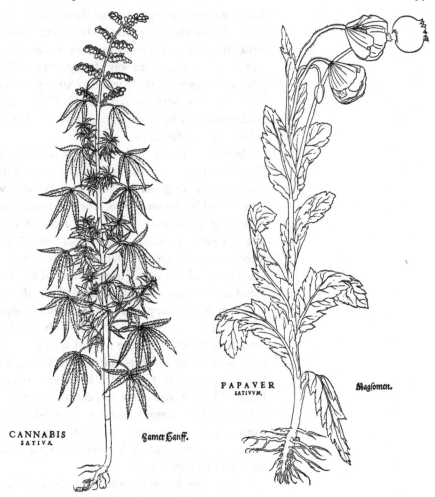

CANNABIS
SATIVA.   Zamer Hanff.

PAPAVER
SATIVVM.   Magsomen.

Fig. 39 *(left)*. Hemp (*Cannabis sativa*). Woodcut illustration from the herbal by Leonhart Fuchs, printed in Basle, *De historia stirpium*, 1542. The '*sativa*' indicates a cultivated plant. It is the female (seed-producing) plant that is shown.

Fig. 40 *(right)*. Opium poppy (*Papaver somniferum*), from Fuchs' Herbal, 1542.

that served as anodynes for many painful conditions, in domestic animals as well as humans.

The Fenlands were tied to this useful but sinister plant by the coincidence that they had been for centuries the home of 'ague', a disease described as 'an endemic, incommunicable, paroxysmal fever', from which there was an extremely high death rate, and the cause of which was subject only to speculation, such that it was due to a 'miasma' carried in 'noxious exhalations' from the vast bodies of stagnant open water. It provoked great misery and fear, especially to the poorest classes condemned to work and live in these conditions. The one generally accepted alleviation lay in use of the opium preparations and it is no wonder that their common employment led to widespread addiction, so that opium-chewing was common, and druggists in the townships such as Wisbech and Holbeach sold great quantities of 'the stuff' across the counter, often as much as 14 lb (6.4 kg) in a day. It was soon recognised that these drugs could reduce not only rebellious stomachs but also fractious children to quiescence, and a Fenland physician writes that 'a patch of white poppies was usually found in most Fen gardens' and that poppy-head tea was often given to children during teething, a risky practice in view of the particular susceptibility of children to morphine. They too were often treated with Godfrey's cordial, or one or two drops of laudanum on a lump of sugar, either providing a very early initiation to habitual use of the drug.

Happily ague diminished greatly with the accelerated drainage of the Fens after 1850. The reason for this was apparent as it became clear that the ague, now more generally known as malaria, was a trypanosomal disease transmitted by the mosquito, *Anopheles maculipennis*, which had its greatest centres of abundance in the two areas of Britain, the Fenland and the Thames estuary, where malaria was endemic. The requirements of the insect's larval stage for stagnant water were fully met in both regions. Drainage, and the use of quinine and of such later drugs as mepacrin and paludrin, have now virtually removed the threat of endemic malaria.

It is generally supposed that the opium poppy was to some extent grown in the Fens as a field crop and certainly the deep, fertile, humic soils that remain calcareous and the overall temperate to warm summers meet the chief requirements for its cultivation. Babington's *Flora of Cambridgeshire* (1860) reports that it is 'said to have been formerly largely cultivated in the Fens'. There is also some indication of former cultivation in a scatter of field- and farm-names such as 'Poppy-Hill', but more precise evidence would be welcome, especially since in *The Survey of Agriculture of the County of Cambridge* by Vancouver (1794) and Gooch (1813) there is no mention of a poppy crop.

It has to be borne in mind that the seed has a high fat content and on extraction yields a marketable oil that is not only edible, but that has use as a drying oil favoured by artists and that was formerly employed in varnish manufacture. The residue after extraction was, as with cole-seed, flax and hemp, a valuable cattle-food. One cannot avoid the speculation that, with three major oil crops already widely grown in the Fenland, it might have been for oil production rather than drugs that field crops of opium poppy were grown, if indeed it was thus cultivated. Whichever it may have been, we may confidently assume that it was competition from areas of cheaper labour and lower production costs or from new products that led to abandonment of the Fen poppy crop.

The woad plant or 'glastrum', *Isatis tinctoria*, was easily the most valuable dye plant of the Middle Ages, not only in this country but throughout Europe. Here a special aura surrounds it, possibly associated with the thought of our painted Iron Age ancestors facing the Romans during their conquest of this country in 54 B.C. Caesar's phraseology is certainly vivid enough, '*omnes vero se brittane vitro inficiunt, quod caeruleum efficit colorem atque hoc horridiores sunt in pugna aspectu*', i.e., 'the Britons stained their skins blue with woad and so took on a wild and fearful appearance in battle'. There is no doubt that woad was the plant mentioned for the culture of *vitrum* and that it was well known in the ancient Greek and Roman cultures. This ancient husbandry of it makes it hard to define its territory as a wild plant, but this was probably southern Russia and the Caucasus. It was possibly the Iron Age people themselves, reaching Britain from the Continent, who introduced it. Numerous historical references testify to the fact that woad cultivation was well-established in this country in Anglo-Saxon time and indeed 'wad' was also imported. All through the Middle Ages, when the wealth of England rested so securely upon its wool production there was an extremely large importation of woad for dyeing it, chiefly from France (along with wine) and the Low Countries, but at the same time so much was now grown at home that there was also substantial export of it, mainly through the Port of London.

The extremely comprehensive book by J. B. Hurry (1930) devoted to the woad plant and its dye makes it superfluous to present more than a brief outline to which one can relate the Fenland history of its cultivation.

The chief centres of woad production in Britain were determined by convenience for the areas of major woollen cloth manufacture and by soils sufficiently deep and fertile for this crop, that like so many cultivated *Cruciferae* is very exhausting of the land. Although subsidiary production was widespread the two chief woad-growing regions were those of Somerset, where the Abbeys of Glastonbury and Muchelney derived great wealth from the crop, and, to a still greater extent, the Fenland of East Anglia. It was the persistence of the woad crop on the silt-fens round Boston, Spalding and Wisbech, when it had ceased everywhere else, that gave special emphasis to the Fenland's association with the crop, and our records, written and pictorial, of the cultivation, harvesting and handling of the crop derive in large part from the few last Fen farms, where cultivation persisted into the early decades of this century.

Woad is a very robust biennial plant, something like mustard and with similar yellow flowers produced in the second year. However, it was generally grown as an annual from spring-sown seed, following deep ploughing earlier in the year. The laborious hand-weeding and singling of the crop was done kneeling, mostly by women, and at quite an early stage of development of the rosette of basal leaves, the first crop was taken by pulling off the fresh green leaves which were carried in large skips to the temporary crushing mills set up in the field or near by. Three or four crops of leaves were pulled during the summer and a second weeding intervened. The mills contained characteristic large wooden rollers, truncately conical that bore sharp wooden or iron slats on their periphery so that as they were rotated by horse-power in the paved circular well, the tipped-in leaves were broken and crushed. The crushed foliage was rolled and compacted by hand and placed upon airy sheltered racks for the spherical lumps about 5 in (13 cm) in diameter to dry. These processes involved very extensive and lasting dyeing of men closely involved in the work, so that they never had the least chance of concealing their trade.

At a suitable stage of drying, probably in the following winter, the balls of woad were crushed in the same mills into a powder which could be packed in barrels for sale to the dyers. Paintings, drawings and photographs of the buildings and processes taken at Parsonage Drove, near Wisbech, fortunately remain.

Woad production was an industry with extremely heavy demand on labour throughout the year, needing as it did, deep cultivation, heavy manuring, repeated weeding and a succession of heavy manual operations in processing the crop. It was thus susceptible to economic pressure so soon as an alternative source of the blue indigo dye became available. This came with arrival of the product of the Far Eastern leguminous plant *Indigofera tinctoria*, and related species, in which the dye was very much more concentrated than in woad. In the reign of Elizabeth I a ban on its import was imposed and this lasted for about a century, but it afterwards entered the country in rapidly increasing amounts and its use, together with that of synthetic indigo, finally sounded the knell of the woad trade, despite the fact that its life was prolonged by employment of woad not as a dye in its own right, but as an essential component in the preparation of the imported indigo and a number of other different dyes in the 'woad vat'. The method led not only to much economy in the amount of dye needed, but some enrichment of colour and better fixation. Even this alternative use had virtually ceased by 1930.

It is difficult to believe that the exchange of the woad crop for bulbs, potatoes, soft fruit and market vegetables of the present-day Marshland has been other than highly beneficial to the region, whatever sentimental regrets we may have for the passing of a crop so closely bound up with England's history.

The modern usage of extending the term 'crop', in the sense of a regular harvest, to animal as well as plant products allows us to mention here two of the main supports of the economy of the undrained Fenland, namely fresh-water fish and water-fowl. Darby's account of the medieval Fenland illustrates by reference after reference the extreme importance of fisheries in the meres and streams, and his extracts from, and maps of, the Domesday returns for Norfolk, Lincolnshire and Cambridgeshire shew great profusion of allocated fisheries and fishing boats, with the yearly eel production alone stated in thousands and amounting to a vast total crop, which was, however, supplemented by a great variety of other fish. Their taking was by the use of nets and fish weirs or *gurgites* along the rivers and Darby uses the telling phrase that 'Eels, indeed, fulfilled many of the uses of currency in the region' and records that debts, rentals and tithes were commonly paid in eels. The great ecclesiastical houses in and around the Fens not surprisingly arranged for such payments to be in Lent and it could be that the original siting of these houses had something to do with the local availability of fish. At a time when the produce of coastal fisheries could not be transported, the great fresh-water fish markets at Fen-margin townships, such as Cambridge, were of particular importance. Although with the progress of Fen reclamation and the disappearance of the meres, wholesale methods of fishing were abandoned, the taking of eels from the fen drains by the use of osier eel-traps and the five-pronged glaive or eel-spear persisted long enough for specimens of these tools to be preserved and even for the making of documentary colour film for television, with one of the last fenmen shewing how they were used. The banks of the larger Fenland waterways are still the magnet drawing thousands of weekend coarse-fish anglers from inland townships as far away as the industrial Midlands.

The fenman who anciently lived off the natural products of the wild Fenland, or so-called 'fen-slodger', depended substantially not only upon fishing, but also on wild-fowling, primarily no doubt for his own household needs, but progressively, as transport improved, for sale to urban centres outside the Fens. It is not difficult to imagine how prolific was the supply, more especially since the considerable breeding population of water-fowl was augmented in the autumn by vast hordes of migratory birds coming from the north continental mainland. There is early evidence of this profusion, as in the twelfth-century *Liber Eliensis*, and in Piercy's *Household Book* of 1512 quoting prices for all Fenland fowl, and in the lists of gifts to supplement an Elizabethan wedding-feast of 1567, that included mallard, teal, swans, cranes, herons, bittern, knots, stints, godwits, curlews, plovers and larks. The smaller birds were in part taken by the use of bird-lime, made from the viscous fruits of mistletoe; as Drayton (1622) has it in his *Polyolbion*

> The toyling *Fisher* here is tewing of his Net:
> The *Fowler* is employed his lymed twigs to set.

Netting was, however, the means preferred to take the smaller water-fowl, such as dunlins, knots, ruffs and reeves, redshanks, lapwings, golden plovers and godwits; these were caught in considerable numbers, especially in suitable localities such as low hillocks in the flooded washes between the major drains, and at the time when Miller wrote of this practice, about 1878, the fowlers were employing tethered decoys and dummies alongside the nets.

With the improvement and more general use of guns, alongside the progress of drainage, fowl became more wary and were most easily taken in secluded and peaceful places, and it was in such sites that there were brought into use, the practice apparently borrowed from Holland, the elaborate techniques of the duck-decoys. These were circular pools from which diverged some eight strongly curved arms, in the entrance to one of which alighting flocks might find shelter whatever the direction of the prevalent wind. The skilful co-operation of the fowler's team of tame decoy ducks and trained 'piper' terrier, excited the curiosity of the wild birds and enticed them deeper down the narrowing arm until they entered the terminal part, netted to form a tunnel. In this, after securing the entrance, they were readily taken. By the early eighteenth century mallard, teal and wigeon were being caught in great quantities, as many as two or three thousand birds a week, for despatch to the London market by carrier wagons from centres such as Peterborough. An observer in 1761 wrote: 'I have often seen these wagons drawn by ten or twelve horses apiece, so heavy were they laden.' Such decoys were numerous throughout the Fenland and traces of them remain, not only in such names as 'Decoy Pool Farm', but in the characteristic star-like depression recognisable in many air-photographs still. One operative decoy at least, persisted into the age of photography, and a few are still kept operative for bird-ringing and for conservation purposes. A powerful addition to the fowler's armoury came with the adoption of the heavy duck-gun which, mounted on a punt propelled silently by the recumbent hunter and aimed by steering the punt, could discharge a lethal spread of shot into a quietly feeding flock of water-birds. For a time around 1700 the considerable income from the decoys was augmented by the lucrative maintenance in the lush Fen meadows, of large flocks of geese, kept primarily for their quills and feathers, often barbarously live-plucked, that were exported in large amount through the port of Boston.

As drainage progressed suitable habitats for water-fowl became more and more restricted. As late as 1786 or so, setting out on foot from Cambridge and going eastwards, one easily reached the fens of Cherryhinton, Teversham, Quy, Bottisham and Swaffham, where a great variety of wild-fowl might be shot. At that time the local lads would furnish the 'University gentlemen' with long poles with which to leap the very wide ditches: an interesting confirmation of the presence and state of the fen drains, but also a reminder of the ancient Fenland (and Dutch) tradition of jumping the ditches by means of such 'Poy sticks'.

By about 1850 the sportsman and growing numbers of scientific naturalists, the two classes by no means exclusive of one another, had to seek further afield for centres of abundant wild quarry. Many of them found a suitable centre in the Lord Nelson Inn by the ferry across the Cam at Upware, better known as the 'Five Miles from Anywhere: no Hurry'. They formed, in the student way, a club with a mock establishment of the 'Upware Republic' and, among much amiable nonsense, many interesting observations on the wild life of the adjoining fens were recorded by its members, a good number of whom subsequently were men of very high scientific repute.

The same area provides a convenient illustration to bridge the gap between the days of the slodger and wild-fowler and the present era when wild fowl gather abundantly only in the refugia that conservationists provide for them. The naturalist artist Dr Eric Ennion has compellingly described the history of the large triangular area of peat-land, the 'Adventurers' Fen' between Wicken and Reach Lodes, and followed its fortunes after coprolite extraction and peat digging had given place to cultivation in the decade 1910–20 that included the 1914-18 war with its severe food shortages. The Fen was subsequently allowed to slip back into a totally waterlogged state, where vast reed-beds and sheets of shallow open water encouraged the return and breeding of a wealth of Fen water-fowl, lovingly observed and portrayed and where a certain amount of organised duck shooting was possible. Re-drained more thoroughly during the Second World War, it has not been allowed a second time to pass out of agricultural use save in the segment still belonging to the National Trust close to the junction of Wicken and Burwell Lodes where a large secluded mere and extensive reed-beds are maintained primarily to encourage the ancient avifauna of the Fens.

16

Lost and vanishing species: conservation

The evidence of increasing rarity or total loss of species of plants and animals once familiar in the Fenland is so well attested by the historical record, and indeed so freshly within the range of present recollection, that a great many people and all fen inhabitants are highly sensitive to it.

The awareness of loss of part of the heritage of our region has always gained a dramatic emphasis from the skeletons of unfamiliar animals dug up in drainage works, peat cutting and agriculture. The most startling of such discoveries include bones of such beasts as the cave lion, cave bear and straight-tusked elephant, that are now totally extinct, and species of hippopotamus and rhinoceros now living precariously and far from Britain. All such derive from the sands and gravels of the Fen islands and the Fen margins and greatly predate the infilling of the Fenland basin after the last Ice Age. Happily the closing stages of that Ice Age itself, the Devensian, are well represented in such river-terrace deposits as those encountered in the recent *Relief Channel* at Wretton, at Earith where the Ouse enters the Fens and at Barnwell Station, excavated somewhat earlier, in the deposits of the next to lowest terrace of the River Cam. In all of these there have been discovered rich floras indicative of cold continental tundra and steppe conditions and a tree-less vegetation rich in arctic plants, dwarf shrubs, sedges, grasses and salt-tolerant herbs, part of a landscape subject to great temperature extremes that threw the surface soils into convolutions and split them with deep frost wedges. Alongside these effects, the high summer temperatures had produced lush vegetation in the innumerable shallow pools, so that the southern English landscape sustained, as the rich fossil fauna shews, large numbers of giant mammalia such as the bison, wild horse, woolly rhinoceros, reindeer, giant Irish deer and mammoth, as well as a rich population of ground-dwelling rodents including species of lemming, voles and shrews.

Only the very end of this cold glacial stage is represented in the Fenland proper, and then in the lowest muds deposited in the deep river channels cut down by the main streams far below their present base-level, during a time when the North Sea basin was itself dry land. By the time of the great marine

Plate 52. A cluster of droppings of the elk (*Alces alces*) found in the aquatic *Sphagnum* peat between the top of the Fen Clay and the overlying shell-marl at Ugg Mere, Huntingdonshire. The droppings are conspicuous because, since the area was drained, they have developed a covering of fungal mycelium. To judge from the pollen analyses and a radiocarbon date of 1310±110 b.c., the animal was therefore alive in the Bronze Age.

invasion that flooded the vast lowland forests of the Fen floor about 2500 B.C., we had advanced a long way into the present Flandrian period and were experiencing mean temperatures above those of today. The late-Glacial open landscape had vanished beneath the mantle of nearly continuous deciduous high forest and a largely different suite of fossil remains is recorded from this time onwards. Some of these we have already made use of to indicate the climatic or environmental conditions prevailing when they were part of the contemporary scene, notably the wood-boring insect, *Cerambyx cerdo* from the buried oak forest and the pond-tortoise, *Emys orbicularis*, both of them good thermal indicators. The basal forest has yielded several specimens of the aurochs, the giant wild ox, including a complete specimen from Burwell Fen besides the already mentioned skull with smashed-in frontal bone, and a pollen-dated skeleton from Mildenhall Fen. This great creature, that was wild in Britain until about the tenth century, was a browsing animal of the forest, where too the fossil evidence points to abundant wild boar, red deer and roebuck with the wolf and the brown bear that also persisted wild in Britain until about A.D. 1100. There is doubt whether the giant Irish deer persisted into the Flandrian Fenland, but a very odd circumstance shews that the elk, *Alces alces*, an animal browsing largely on shrubs and submerged water-plants, persisted at least into the Iron Age. In looking into the stratigraphy of the drained Ugg Mere, I had found in aquatic *Sphagnum* peat below the shell-marl, what looked like a nest of about thirty 'eggs' of brown *Sphagnum* moss somewhat pressed together, each about the size of a thrush's egg and each separately covered with a thin white weft of fungi (Plate 52). Not until I had visited Finland many years later did I realise that this was a cluster of elk droppings, such as one finds on the wet shore of shallow lakes in which the animal regularly grazes. Carbon-dating confirmed its age at about 1300 b.c. and a similar discovery of a little younger age has been made also in raised-bog peat in Somerset.

An obvious explanation of the loss of these giant mammals from the present fauna is the loss of a suitable environment with destruction first of the cold-stage grazing and then of the continuous high forest, but it does not do to forget that these large beasts have always been relentlessly hunted by man; there is excellent fossil evidence that late-glacial Mesolithic man hunted the elk in Lancashire, and one of the world's oldest wooden weapons is a pointed and fire-hardened yew wood spear found in the coastal deposits of Essex thrust between the ribs of an interglacial elephant. The American scientist P. S. Martin goes so far as to attribute the world-wide extinction of the great Pleistocene megafauna almost wholly to human 'Overkill'.

It could well be that man also had some hand in the extinction from Britain of the beaver, *Castor fiber*, doomed of course in any event by progressive fen-drainage. Several skeletons of the animals are known from various parts of the Fenland: the skeleton shown in Plate *53* and the lower jaw drawn by Ennion from Burwell Fen are safely attributable to the peat deposits and of course water and timber were the beaver's prime requirements: Fowler reported one discovery of a beaver dam. The otter,

Plate *53*. Skeleton of the beaver, *Castor fiber*, recovered from the peat in Burwell Fen, Cambs. Note the enormous incisors, used in felling small timber, the very wide chest and the powerful back feet and tail. Vertical scale about ⅓. Sedgwick Museum, Cambridge.

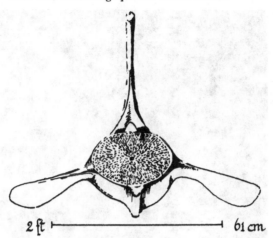

Fig. 41. Vertebra of a whale from the fen deposits (probably the roddon silts) at Littleport, 1961.

2 ft   ⊢————————————————⊣   61 cm

that still persists here and there in the Fenland mires is, as one would expect, known also fossil.

No category of Fenland animal has been more directly affected by the post-Vermuyden drainage than the water-fowl. To the historical evidence for this, the fossil discoveries add records of the pelican. There had been four previous East Anglian records of individual bones, when the exposures in the *Relief Channel* provided a further record for which pollen analysis allowed attribution to the Iron Age: it was not possible to refer the bird specifically to either the Dalmatian or the White Pelican, the two species at present living in Europe. The pelican was found in the pre-Roman Iron Age lake village settlement at Glastonbury, but there is no clear evidence that the bird was native here in Roman or later times, although stray individuals from time to time reach this country. For other water-birds now rare or absent from the Fens, such as the crane or bittern, the fossil record does no more than establish the prehistoric end of a story that is far more fully known from direct historical accounts and eye-witness testimony.

In Chapters 7 and 9 where the origins respectively of the Fen Clay and the Upper Silts and Clays were considered, very extensive use was made of the fossil content of the beds as evidence for the salinity of the water in which they were laid down. We owe much more in a geological sense to the vast numbers of microscopic foraminifera, and to a smaller degree the diatoms, whose shells typify these deposits than to the very infrequent, but still dramatic discoveries of such vast marine vertebrates as the walrus whose skull came from the Ely fens, the two 14 ft (4.25 m) whales, *Pseudorca crassidens*, reported in 1920 from Bassenhally Fen or the scattered occurrences of massive whale vertebrae (see Fig. 41). Dr Macfadyen grouped the Fenland fossil foraminifera, essentially marine organisms, into six categories of tolerance of brackish-water, and the relative abundance of

Plate 54. The sundew, *Drosera rotundifolia*, a vanished Fenland plant typical of acidic heaths and *Sphagnum* bogs. It has a basal rosette of leaves, each with a radial display of tentacles, all of which carry a glistening drop of sticky fluid. An alighting insect, held by the fluid, stimulates the tentacles that quickly enfold it until digestion of the prey is completed. (Photograph by M. C. F. Proctor.)

Plate 55. The cranberry, *Vaccinium oxycoccus*, in fruit. Its fine wiry red stems trail over the growing shoots of *Sphagnum* moss, and slender arching stems support the brilliant red berries just above the bog surface. The small leathery green leaves are visible at the top of the picture, which is about life-size. (Photograph by M. C. F. Proctor.)

tolerant and intolerant forms allowed him, as we have noted, to distinguish sharply between the salinity of the brackish Fen Clay lagoons and the far more marine environment of the tidal waters in which the Roddon and Marshland silts and clays were deposited. Relatively seldom are marine fossils of intermediate size between the foraminifera and the cetaceans recorded, but the mollusc *Scrobicularia piperita* is fairly common and now and again, as at Glass Moor or in the bed of Whittlesey Mere, one finds the abundant shells of cockles (*Cardium edule*) that briefly lived on the surface of the peat or on the stumps of Fen trees when the earliest sea-flooding of the Fen Clay transgression reached the region. It is so long since the last effective marine inundation of the Fenland that these salt-water organisms now no longer occur at all in the peat areas, save where some large stream such as the New Bedford River is kept tidal.

A similar but far more recent parallel to the loss of the marine and brackish-water organisms lies in the hitherto unsuspected effects of destruction by peat-cutting and drainage of all traces of the acidic raised bogs that formerly occurred round the Fenland margins, especially as we have seen in the Holme Fen–Whittlesey area, but conjecturally far wider. The tardiness of recognising this effect is not surprising when we recall that only a hundred years ago even the idea of the presence of acidic bog plants was dismissed summarily, Miller contradicting outright the statement by S. Wells (1830) that 'the turf moors are covered with such plants as the Heath, Ling and Fern, the *Myrica gale* . . . and a grass with a beautiful white tuft called the Cotton Grass'.

The limited success of conversion of Holme Fen to arable cultivation and its subsequent determined preservation for game, in fact ensured the tenacious persistence of a handful of these very species down to the present day. In the interim there had been a thin scatter of records by Victorian botanists, among whom the Countess of Huntly incorporated in her herbarium a labelled specimen of *Drosera*, the sundew, whose sensitive sticky tentacles enfold trapped insects until digestion by the leaf has been completed (Plate 54). Later still, Cambridge botanists in many field excursions and investigations have been able to identify from the acidic peat itself, first at Wood Fen and then at Woodwalton, Holme and Trundle Mere, a very long list of plants highly typical of raised bogs. These naturally include many species of the peat-forming *Sphagnum* together with associated and highly typical species of other mosses and liverworts. The remains of flowering plants include the ling, bell-heather, cross-leaved heath and the less familiar *Andromeda polifolia* (typical of the margins of bog pools), the surface-creeping cranberry, *Vaccinium oxycoccus* (Plates 55 & 56), with the taller bilberry, *V. myrtillus* and the cowberry, *V. vitis-idaea*, together with the shrubby crowberry, *Empetrum nigrum*, and, reassuringly,

Plate *56*. The erect flowering shoot of the cranberry, rising some 2 to 3 cm (1 in) above the wet *Sphagnum* tussocks. This pedicel, like the trailing stems, carries the glossy dark green, leathery leaves. This plant, once harvested on the acidic Fenland, has vanished with the disappearance of the old *Sphagnum* bogs. (Photograph by M. C. F. Proctor.)

even the sundews. Among identified monocotyledons are the bog-asphodel, *Narthecium ossifragum*, the white beaked-sedge, *Rhyncospora alba*, and the two commoner species of cotton-grass, together with other typical but less easily recognised species of the sedges. There have even been identified the narrow rhizomes, clad in their papery leaf-sheaths, of a bog-plant, *Scheuchzeria palustris*, now all but extinct in the British Isles (Plate *27*).

Since the raised bogs naturally formed at the top of the peat sequence, they were particularly susceptible to drainage and peat-cutting, whilst in their half-dried condition they were subject to almost instant colonisation by birch and to devastation by fire. Their loss has quite certainly involved the Fenland extinction of many animals, especially the insects that are specialised to them. Thus there still exists a museum specimen taken long ago from Wicken Fen, of *Trichoptilus paludum*, an insect that lives exclusively upon the sticky glands of the sundew leaf. Less decisively one notices that in description of Whittlesey Mere shortly before its drainage, the shores are reported as the home of large numbers of the common adder. My own experience is that this snake is generally common on the hummocky *Sphagnum*-clad surface of raised bogs, such as we know came down to the edge of the mere. The viper is now exceedingly rare throughout the peat Fens, not altogether surprisingly: it is by no means pleasant suddenly to find a viper sunning himself on the stack of turves where you were about to put your hand, so the older fenmen killed the snakes wherever opportunity offered, as peat-diggers do everywhere in Britain. It is interesting that Dr E. Duffey reports from Woodwalton Fen the persistence of a single species of spider, *Lycosa paludicola*, that is unequivocally associated with the acidic bog that we now are certain formerly existed there.

I think we must acknowledge that in the living raised bogs of the Fenland there was indeed a treasury of wild life, lost to us almost before their presence had been recognised.

A totally different but still appropriate instance of the way in which altered conditions have caused the disappearance of a Fenland plant, is the loss of the broomrape, *Orobanche ramosa*. This is a totally parasitic plant, devoid of chlorophyll, that grows exclusively on the roots of the hemp; it was recorded in the late eighteenth century as frequent in the hemp fields about Wisbech and to a less extent was known from near Upware, but it has gone altogether with the *Cannabis* crop. On the whole the weeds of cultivation in the Fens have had a history not greatly different from those of other arable lands, though ditch-side aquatics have often in wet seasons strongly supplemented the general run.

It is therefore much as we might expect, the category of wild plants that has suffered severest restriction and extinction in consequence of Fen drainage and exploitation has been that of species of open water, reed-swamp and wet fen, that is to say the natural components of the early stages of the hydrarch vegetational succession. For many such plants we were in time to have authentic records: others persist in a few localities still. Among the plants of open water we may mention the water-soldier, *Stratiotes aloides*, whose rosettes of narrow serrated leaves float below the water surface linked together by thin rhizomes: as they accumulate diatoms and precipitated carbonate they sink to the mud bottom, recovering in spring from small perennating shoots or 'turions' (Fig. 42). Oddly enough, though the plants are individually male and female, as a rule only the latter now grow in Britain. This handsome enigmatic plant with a long fossil history, once widespread and found at Wicken Fen, recently has been restricted to one remaining Fenland site. Two plants becoming much less

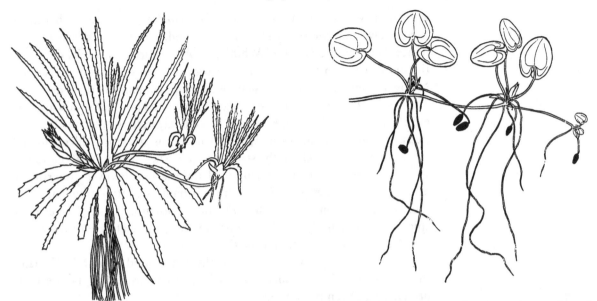

Fig. 42 *(above)*. The sub-merged aquatic plant, the water-soldier (*Stratiotes aloides*) now a rare Fenland species: the plantlets, borne on side shoots after flowering, propagate the plant vegetati-vely. From A. Arber.

Fig. 43 *(right)*. The frog-bit (*Hydrocharis morsus-ranae*) whose small glossy floating leaves used commonly to cover the Fen ditches. Note the small winter-buds (shown black) that are detached, sink to the bottom, germinating in the spring and rising again to the water surface. From A. Arber.

common also, are the white-flowered frog-bit with glossy heart-shaped leaves floating like miniature water-lilies in the quiet fen-ditches, and the somewhat larger *Limnanthemum peltatum*, that has a very similar habit though a good deal larger, and beautifully fringed yellow flowers recalling those of the bog-bean, *Menyanthes trifoliata*, that also belongs to the family Gentianaceae. There is no reason to doubt the native status of *Limnanthe-mum*, for all that it chiefly occurs in the big man-made drains and has left no fossil evidence of its earlier presence.

It seems likely that many of the larger water-side or wet-fen species have been particularly at risk by drainage and mowing. Thus three large yellow composites were once frequent in such Fen situations, the tallest, that might attain 9 or 10 ft (3 m) in height, was the marsh sow-thistle, *Sonchus palustris*, recorded as late as 1763, then extinct but re-introduced in Woodwalton Fen, and since found scattered in a few other Fen localities. A second somewhat less imposing plant, attaining however some 3 ft in height (1 m), is the marsh fleawort, *Senecio palustris*, recorded by John Ray in 1660 and known in the Fens for perhaps one hundred and seventy years afterwards: it now seems to be extinct.

The great fen ragwort, *Senecio paludosus*, about 2 m tall at most, was fairly often recorded in the Fenland ditches from the seventeenth century onwards, but by about 1865 British botanists were reconciled to its apparent extinction from the flora of this country. In 1972, however, a small colony of five plants was discovered within the area of its former range, and it has yielded fertile seed now happily grown on to flowering in the Cambridge Botanic Garden (Plate 57). The possibility of seed retaining viability in

Plate 57. The great fen ragwort (*Senecio paludosus*) raised in the Cambridge Botanic Garden from fruit gathered from the wild plant re-discovered in 1972 after an absence from the Fens of some one hundred and fifteen years. (Photograph by W. H. Palmer.)

moist soil for some centuries is better regarded now than formerly and it seems possible that the colony came from germination of long-buried achenes in the re-dug ditch where it occurs.

Emergent water-plants of this large stature sometimes play a critical rôle in the survival of fen insects for which they are important or obligatory hosts. Thus caterpillars of the large Swallow-tail butterfly, *Papilio machaon* (Plates 58 & 59), that used formerly to occur in great numbers round Whittlesey Mere and elsewhere throughout the Fens, feed exclusively upon

Plate *58*. Caterpillar of the Swallow-tail butterfly feeding on its commonest host plant, the milk-parsley (*Peucedanum palustre*) at Wicken Fen. (Photograph by W. H. Palmer.)

Plate *59*. Adult Swallow-tail butterfly (*Papilio machaon*), a species for which Wicken Fen was long a famous locality. It died out some twenty or thirty years ago, and this photograph and that above, though taken at Wicken in 1969, are of the Norfolk race recently introduced in an effort at re-establishment of the species. (Photograph by W. H. Palmer.)

the Milk Parsley, *Peucedanum palustre*, a robust umbellifer of mixed-sedge communities. The decline and extinction about 1949 of the Swallow-tail from Wicken Fen, formerly noted for it, is generally associated with great decrease in abundance of the host-plant and it has been accepted, after one or two abortive attempts at re-introduction, that any such attempt, to be successful must be preceded by substantial re-establishment of *Peucedanum*. Similarly the Large Copper butterfly, *Lycaena dispar*, depends upon the giant water-dock, *Rumex hydrolapathum*, a plant that, since re-introduction, flourishes at Woodwalton Fen, once famous for this insect and where, the native strain having disappeared, a Dutch strain has been successfully established in place of it. Other plants of similar habitat suffering diminution are the giant spearwort, *Ranunculus lingua*, the yellow loosestrife, *Lysimachia vulgaris*, and the dwarf willow *Salix repens*, var. *fusca*; the last two are strongly associated with the mixed sedge crop at Wicken Fen, the first especially with its derelict boat dykes. Only plants of large size and vigorous habit can commonly withstand competition of the tall evergreen *Cladium* leaves and their dead remains, but one common associate, the marsh pea *Lathyrus palustris*, has slender scrambling stems that grow from thin perennial stems below ground. It formerly occurred widely in the Fens but is now restricted to Wicken (Plate 60); its sites elsewhere in the country are few and diminishing.

Not all the lost or vanishing fen species belong, however, to this size category or to the *Cladietum*. The small fen orchid, *Liparis loesellii*, was

Plate 60. Two plants strongly associated with the 'mixed-sedge' crop. The robust yellow loosestrife (*Lysimachia vulgaris*) and the scrambling marsh-pea (*Lathyrus palustris*). Wicken Fen, August 1974. (Photograph by W. H. Palmer.)

Plate *61*. The yellow water-lily or 'brandy-bottle', *Nuphar luteum* and emergent flowering stems of mare's-tail, *Hippuris vulgaris*. In the clear water of Wicken Lode, as it was in 1957, the underwater leaves of the water-lily are easily visible. Since the Lode has become turbid through excessive boating, the lily has greatly diminished and the mare's-tail has apparently vanished. (Photograph by W. H. Palmer.)

recorded from the time of John Ray's seventeenth-century conspectus, and Babington's 1860 *Flora of Cambridgeshire*, as abundant in the fens near Cambridge, but it seems last to have been seen at Wicken Fen in 1945: it needed the very wet moss communities provided by intermittent small-scale peat diggings such as formerly obtained in the 'Poor's Pieces', and it has vanished with cessation of peat-cutting. It seems more than likely that a similar cause explains the disappearance of the handsome Fen Violet, *Viola stagnina*, once abundant at Wicken where it is now extinct, although it persists at Woodwalton Fen.

Such observations as these strengthen the conclusions of authorities on other groups of organisms, for example the spiders and mollusca, that an extremely important rôle in the extinction of both plants and animals has been played by changes in Fen usage and practices of Fen exploitation that have been larger in scale and more decisive in character than is generally suspected. In the last few years it has seemed increasingly likely to me that a number of the puzzling features of the plant-life of Wicken Fen may find explanation in effects of this kind. In particular we have the remarkable past records of such acidic bog plants as sundew, *Drosera* sp., and cotton-grass, *Eriophorum angustifolium*, coupled with the present-day persistence of patches of sweet-gale, *Myrica gale*, and a remarkable abundance of the alder-buckthorn, *Frangula alnus*. The two latter species are far more characteristic of acidic heathy soils and moderately acidic peats than of calcareous fen, and indeed often grow freely in the acidic fens of the low drainage channels (lagg-streams) that separate and limit the edges of raised bogs. Such sites must in fact have bordered any raised bogs that formerly existed at Wicken, and could have been the source from which these two shrubs re-established themselves after a phase of the very extensive peat-cutting to which repeated air surveys now bear very strong testimony. Contrary as it is to the habit of some decades to refer to Wicken Fen as a 'relic of the primitive fen' and so forth, it is at least worth conjecture whether some feet of top peat, including those marginal domed areas that had become acidic, have not been totally cut away, and left to re-colonisation from the extensive lagg-streams and drainage channels left undisturbed. To encourage this speculation we have the surprising fact noted in our earliest investigations of the peat, that though all the remaining peat is full of the pollen, wood and cones of the alder, as would be expected, there were then and for many years, no living alders on the Fen surface save one or two planted trees. Even the sword-sedge itself, so much the hall-mark of Wicken's economy and vegetation over half a century at least, is a plant that, as A. Young noted in his surveys of *c.* 1800, thrives particularly well in abandoned peat-cuttings. Of course so vast an assemblage of species of aquatic and sub-aquatic plants and animals of every kind, must inevitably have persisted locally through any such period of extensive peat-removal, but none the less a hypothesis of the kind outlined must now be taken seriously into account, as regards both understanding of the fen's ecology and its best future management.

CONSERVATION

Our account will have shewn how greatly loss of fen habitats and destruction of native flora and fauna accelerated from the mid nineteenth century when the last great meres were drained and cultivated. Through the second half of that century botanists and zoologists were still mainly collectors and classifiers, often with very extensive field-knowledge, but this generally directed towards more exact description and identification. It is interesting that not until 1895 did the Cambridge University Chair of Botany come to be occupied by someone *not* primarily an ardent taxonomist or field botanist. By now, however, all field biologists had come to be affected by the growth of other natural sciences, especially by geology and by knowledge of the physics and chemistry of the natural environment, effects quite evident in the highly creative work of Charles Darwin. From these interactions there arose spontaneously in different countries of the world the ideas and doctrines that in the present century have crystallised into the science of ecology.

It is against this background that we have to view the setting-up of the

Plate 62. The Main Drove, Wicken Fen, looking east towards Upware: 1928. The drove, here shewn recently scythed, was cut two or three times a year so that dwarfer plants were never excluded by competition and were present in variety. This helped to make it a favourite collecting ground for lepidopterists, who used treacle-smeared cork nailed to posts (seen right) to attract them. The strip of Fen carrying the black hut and the birch trees had already been uncut for many years and was already covered with carr.

early nature-reserves. With foresight and great sense of public responsibility, a strip of land in Wicken Fen was given in 1899 by J. C. Moberley to the National Trust, to be followed extensively by large numbers of individual donors, so that today 730 acres (300 hectares) are in the hands of the Trust. Many of the donors were extremely able amateur entomologist collectors, such as G. H. Verrall who presented a very substantial area in 1911, and they were clearly concerned to preserve the extraordinary wealth of insect life for which the Fen was notable. In the period of this acquisition and establishment, on suitable evenings moths were collected at stations down the Main Drove by the use of brilliantly-lit white sheets, whilst 'treacle'-smeared cork nailed to posts was an alternative attraction. The avidity of collecting about 1885 is shewn by the fact that the cottage at the entry to the drove rented by the Farrens, father and son, was called 'Catch 'em all', by the Farrens' pride in their annual collection figures and by references to professional 'sedge-cutters cum moth-collectors' in the village. The seemingly inexhaustible riches of the Fen stimulated similar collecting of most groups of plants and animals, and it has given us a remarkable store of records and reference material. It was soon realised by the management committee that their task amounted to much more than controlling access, preventing disturbance and simply 'leaving nature to itself'. With the acceptance of an ecological attitude it was understood also that reserves are far more than outdoor living museums where the specimens helpfully curate themselves. They are rather to be regarded as open-air laboratories where scientific observation and experiment upon wild species and natural communities can be pursued with security from interruption and with suitable facilities both for research and education upon all the growing body of knowledge of the relationships between the plants and animals and their habitat.

As there emerged the recognition of the reality of primary vegetational succession on the Fen, it was evident that at bottom there was little that could be done on a broad scale to halt the progressive effects of peat-growth and the consequent elimination of the more aquatic communities or the general progress of drying out of the Fen. This realisation led the Committee to encourage the more aquatic type of plant and animal life of the Fen by digging out first a small and secluded decoy pond, and then a large open-water mere that since 1955 has grown large reed beds and has attracted thousands of water-birds to Adventurers' Fen. More lately a group of shallower excavations has supplemented the large mere. The Committee has accepted that if you cannot bring the water-level up to ground surface, then you must dig ground surface down to water-level!

However important it proved to agree this principle, when plant ecologists came to interpret the character of the extensive areas of Wicken that were covered by the herbaceous communities that harboured much of the typical plant and animal life of the Fen, they found it difficult to explain them merely in terms of the natural succession. It was S. M. Wadham (later Professor Sir Samuel Wadham, of Melbourne) who first realised that they were probably associated with the traditional cropping of the Fen to which we have referred. It became clear on investigation that taking a crop of the sword-sedge every three to five years, or of grass- or rush-dominated 'litter'

Plate *63*. Entrance to the Main Drove, Wicken Fen in high summer, 1974: the 'Catch 'em all' cottage used to stand just on the left. On the Fen proper in the middle distance the great encroachment of bushes since 1928 is evident. (Photograph by W. H. Palmer.)

annually, did not *arrest* the main reaction of the primary succession; peat continued to form and the ground to get higher above water-level and thus drier, but the species responding to such changes were restricted to those that could tolerate this cutting. This was formulated as the principle of 'deflected' succession (Fig. 44). Since trees and shrubs were thus excluded, the slowly-rising Fen surface now carried a wealth of less tall herbaceous plants and with this an increasing variety of animal life. Thus originated the very characteristic 'mixed sedge' communities (Fig. 38), but where cutting was annual or biennial the *Cladium* itself was killed, and the competition of this giant evergreen sedge with its persistent mattress of dead leaves once removed, the tussock-forming grass *Molinia coerulea*, with its basal storage shoots unharmed by winter cropping, spread rapidly and with it a host of dwarfer herbaceous species that flourished in the more open vegetation, including for instance the marsh pennywort at ground-level, the devil's bit scabious, the uncommon meadow thistle, fen orchids and two or three species of rush and dwarf sedge (Fig. 45). Where cutting was still more frequent, as along the Main Drove, the vegetation was yet dwarfer and still more various and rich in species both of plants and animals: no wonder it was so favoured a place for collectors.

Of course, because the succession had not been arrested, but only 'deflected', the communities produced by cropping became progressively

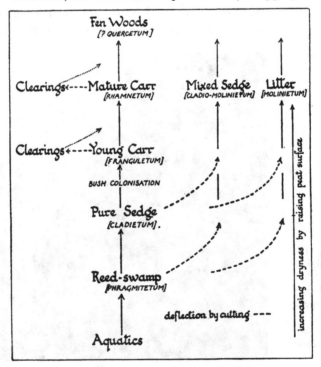

Fig. 44. Scheme showing the relationship between the natural succession of fen vegetation, in response to peat accumulation and rising ground-level (primary succession) to the successions determined by crop cutting (deflected successions) as exhibited at Wicken Fen, but of general applicability.

Fig. 45. Diagrammatic bisect through the 'litter' community at Wicken Fen, a vegetation type induced by yearly crop cutting. The general dominant is the grass *Molinia coerulea* (M) that forms low tussocks full of the over-wintering swollen stem-bases unharmed by winter scything. It is low enough to be accompanied by many other herbaceous plants such as the rush *Juncus obtusiflorus* (J), carnation sedge *Carex panicea* (C), devil's bit scabious *Succisa pratensis* (S.p.), the reed *Phragmites communis* (P), and pennywort *Hydrocotyle vulgaris* (H). Since the 1914–18 war demand for litter and rough chaff has disappeared and the vegetation type is now scarce although the regularly cut droves exhibit some of its character.

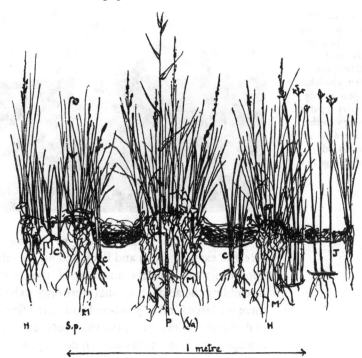

and intensely susceptible to invasion by bushes and trees, the moment cutting ceased. Such invasion was naturally disastrous for all the plant and animal species so typical of the open, cropped, communities. Thus although after the 1914–18 war, crop-cutting as an end in itself had almost ceased, the practice was resumed by the management committee as a means of holding tree and bush colonisation at bay. At first rather tentative and supplemented by a continuing if meagre demand for sedge as thatch, it has of recent years been more rigorously pursued, although a substantial part of the Fen must be allowed to progress swiftly towards fen carr or fen woodland.

Anything but active management is out of the question and this evidently must rest upon as good scientific knowledge as can be acquired. This is borne in upon the management committee even more strongly in the latest few years when the natural drying out of the fen succession has been severely reinforced by a lowering of lode-levels, especially resulting in unprecedented freedom from winter flooding of the Fen. The consequent acidification of the surface peat and the invasion of the shady ground flora of the carr by sixty or more species of acidicolous mosses, liverworts and ferns new to Wicken, introduces a fresh and very important ecological consideration. It presents a dramatic challenge, and one not simplified by strongly accelerated spread of trees, especially the birch. On the one hand a

Plate *64*. Wicken Lode, 1964: view eastwards from the bird-observation hide. The difference in level between the Sedge Fen itself (left) and Adventurers' Fen, which has been subject to peat-cutting and drainage, is very apparent. Even in this winter view the tremendous extent of bush-colonisation (mainly alder-buckthorn) on the Sedge Fen is evident.

development of this scale and kind may be irreversible, despite its alarming threat to existing fauna and flora, and on the other it presumably offers the opportunity to witness and record the re-development of something resembling the acidic bogs that were once to be found in the Fens. Happily there is a large fund of good-will and scientific experience to be drawn upon in deciding and executing the most profitable and practicable adaptation of management to meet this sudden development.

What appears to have happened at Wicken Fen over the last five thousand years or so is broadly as follows. General waterlogging induced by rising sea-level caused fresh-water fen to replace the original high forest, the tall trees of which were entombed by the accumulating peat. Sedge-fen with some alder and willow occupied the ground until, in all probability its surface reached 12 ft (4 m) or more above sea-level, a progress interrupted by local formation of shallow pools forming shell-marl during the Romano-British period. Presumably after this time the natural succession led to the formation of fen and in parts at least of the area, of raised bog. The acidic peat so produced has been now removed by the turf-cutting that may well have operated in medieval time but became very extensive and swift in the seventeenth and eighteenth centuries and especially in the nineteenth century as improved drainage and enclosure awards encouraged it. As the turbaries were exhausted and abandoned to fill with the highly calcareous flood-waters from the upland, the cuttings were massively invaded by fen vegetation from the rivers, drains and marginal fens that always surrounded the acid bogs. Above all there was vast re-expansion of the giant sword-sedge *Cladium mariscus* that became dominant over a large extent of the old turf-cuttings, and became available as a valuable crop for thatch, to exploit which both droves and boat-dykes were cut through the Fen. More frequent cutting than that tolerated by the sedge led to the development of plant communities yielding litter. The growth of peat continued rapidly in these conditions, the peat trenches filled in and the balks between them rounded off so that they are now easily overlooked, and the natural succession has again reached the point of massive bush and tree invasion, pioneered especially by the alder-buckthorn, a relic, like the sweet-gale, of the earlier acidic peat phase. Finally, through the general lowering of water-levels under external control at this critical stage of peat accumulation, it appears that a re-establishment of acid bog may be in progress. There is small wonder that with so varied a history this Fen should harbour so great a variety of plant and animal life, though we may no longer regard it as 'an undisturbed remnant of the primeval Fenland'. It would almost certainly appear on close examination that all the other Fenland conservation areas have suffered similar complex histories, knowledge of which is essential to their proper management.

The second oldest of the larger Fenland Nature Reserves, like Wicken Fen, also originated in a private benefaction, that of the Hon. Charles Rothschild, whose generosity from 1919 onwards enabled the Society for the Promotion of Nature Reserves to acquire over 500 acres (200 hectares) in Woodwalton Fen, Huntingdonshire. This included most of the raised bog land that Sybil Marshall's father recollected as having helped to cut for peat when he was a boy, and that at its northerly end took in the basin of the former Brick Mere. With the vicissitudes of the peat-cutting, some sporadic grazing and hay-making and progressive re-colonisation by scrub, chiefly sallows, there was a wide variety of habitat and survival of some notable fen rarities already mentioned. In 1954 the Society leased the Reserve to the recently founded Nature Conservancy that had come into existence in 1949 in response to very widespread demand that the country should itself provide a central organisation for the acquisition and maintenance of Nature Reserves and for such scientific enquiry as is necessary to manage them. The effective scientific research done in the Fen already, including that of the future director of the Conservancy whilst a Ph.D. student, was extended and a programme of controlled operation of dyke sluices returned some wetness to the over-drained Fen. As a result of this careful management some fen species have reappeared and other rarities are holding their own. Many Fenland plants once present were re-introduced in the 1920s from other parts of Britain and even from the Continent: they are mostly well established but do not necessarily represent the original races.

Near to the Woodwalton Reserve lies that of Holme Fen, an area to which we have already made such extensive reference in earlier pages that it is apparent of how much historic vegetational interest and indeed, geological interest, its preservation must be. When the adjacent Whittlesey Mere had been drained in 1851, and attempts to reclaim Holme Fen for agriculture had been abandoned, the swiftly-drying scarred surface of the acid bog was heavily planted with birch, pine and other trees and when these were clear-felled in the Second World War, natural colonisation by birch rapidly overwhelmed the area. A handful only of scattered relict bog plants was now to be found, and the main railway line running along the margin presented a fire hazard unhappily realised in 1976. In 1952 the area of 640 acres (260 hectares) was acquired by the Nature Conservancy who have since managed it so as to restore, if that is possible still, something of its earlier and intensely characteristic aspect. They are helped by the chance that an early decoy pool, used formerly for taking of duck, has preserved a single damp refuge for a nucleus of a few of the more desirable species of plants.

Successful and esteemed as the Nature Conservancy and its successor, the Nature Conservancy Council, have been, it would be mistaken to suppose that they could be responsible for more than a carefully chosen group of reserves of the highest national priority. There remain great numbers of sites throughout the country, generally of smaller size and of more immediately local and specialised interest (though often of high scientific and educational value) that are more appropriately cared for by local organisations, who are better placed to see the need, to anticipate the threat and secure continuous and watchful oversight of them. Happily the Fenlands share in these more local provisions, those of the Southern Fenland co-ordinated in the active *Cambridgeshire and Isle of Ely Naturalists' Trust*. Whilst there is no room to write of the many local reserves they sustain for residual fen flora and fauna, it is worth while finally to mention their joint activity with the Huntingdonshire and Bedfordshire Naturalists' Trust and the two older national organisations, *The Royal Society for the Protection of Birds* and the *Wildfowl Trust*, in setting up between the Old and New Bedford Rivers what are collectively called the *Ouse Wash Reserves*. Since 1964 these bodies have bought a string of properties in the Washes, extending discontinuously from about 2.5 miles (4 km) north of Welney, to about the same distance north of Mepal, covering some 9.5 miles (15 km) in all, and totalling over 1400 acres (570 hectares). The washes are fairly regularly flooded through the winter and provide lush grazing for stock between April and November, the cattle confined by the low-level drains. The water-meadows provide an uncommonly attractive breeding habitat for water-fowl, especially for familiar waders that are often present in great numbers, and the black-tailed godwit, ruffs and even black terns which occur in lower frequency. Other regular and generally uncommon breeders are the teal, gargeney, pintail, gadwall, shoveller and shelduck. Even more impressive is the regular response to winter flooding, and since the Washes lie right within the great southward migration route from Scandinavia and Russia, vast flocks of wild fowl winter here, including numerous species that favour deeper water than the waders. Though wild

Plate *65*. Sedge-warbler with nestlings. (Photograph by W. S. Farren.)

geese are uncommon, the numbers of Bewick's swans are so great that they are estimated to represent some 10 per cent of the north-west European total population, whilst the pintail and wigeon also represent large fractions of their corresponding totals. One cannot dispute that the Reserve now is one of the most important wild fowl refugia throughout north-west Europe, and it offers a grand illustration of effective preservation of an important segment of the wild life of the ancient Fenland.

The voluntary organisations collaborate amicably in administering the reserve, and their joint management committee deals understandingly with policies of flooding, shooting, grazing and angling so that traditional fenland sports and farming continue harmoniously within the requirements of conservation. Though bird interests properly prevail, the many ditches and flooded meadows provide habitats in which many of the aquatic plants of Fenland prosper, notably sixteen species of the pondweeds, *Potamogeton*, the frog-bit, *Hydrocharis morsus-ranae*, and the fringed water-lily, *Limnanthemum peltatum* that we mentioned among our threatened species.

Possibly the most encouraging aspect of all in the conservation of the wild life of the Fenland is the vast extent of public interest and appreciation: at a recent winter visit to the Ouse Washes, car parking proved next to impossible and the water fowl themselves appeared to be outnumbered. To marshal such enthusiasm is a welcome task to conservationists.

References to general accounts of the Fenland

Astbury, A. K. 1958. *The Black Fens*. Golden Head Press, Cambridge, reprinted E.P. Publishing, Wakefield, 1970.

Bloom, A. 1953. *The Fens*. R. Hale, London.

Darby, H. C. 1940, 1956. *The Draining of the Fens*. Cambridge University Press.

Darby, H. C. 1940. *The Medieval Fenland*. Cambridge University Press, reprinted David & Charles, Newton Abbot, 1974.

Ennion, E. A. R. 1942, 1949. *Adventurers Fen*. Herbert Jenkins, London.

Fox, C. 1923. *The Archaeology of the Cambridge Region*. Cambridge University Press.

Grove, R. 1976. *The Cambridgeshire Coprolite Rush*. Oleander Press, Cambridge.

Hurry, J. B. 1930. *The Woad Plant and its Dye*. Oxford University Press.

Marshall, S. 1967. *Fenland Chronicle*. Cambridge University Press.

Miller, S. H. & Skertchly, S. B. J. 1878. *The Fenland Past and Present*. Leach & Son., Wisbech.

Parker, A. K. & Pye, D. 1976. *The Fenland*. David & Charles, Newton Abbot.

Porter, E. M. 1969. *Cambridgeshire Customs and Folklore*. Routledge & Kegan Paul, London.

Ravensdale, J. R. 1973. *Liable to Floods*. Cambridge University Press.

Salway, P. *et al.*, ed. Phillips, C. W. 1970. *The Fenland in Roman Times*. Roy. Geog. Soc., Research Memoir, 5.

Skertchly, S. B. J. 1877. *The Geology of the Fenland*. Memoirs Geol. Survey England and Wales.

Soil Survey of Great Britain: Memoir. 1974. *Soils of the Ely District*. Sheet 173.

Wilson, J. K. 1972. *Fenland Barge Traffic*. R. A. Wilson, Kettering.

Readers who want to consult the original sources of material used in this book will be able to trace them in articles under the name of the author, Fenland Research Committee colleagues and others referred to in the text, that have appeared in specialised journals such as the *Journal of Ecology*, the *Philosophical Transactions of the Royal Society*, the *Antiquaries Journal* and *Geographical Journal* or in references cited in these papers.

Index

Bold type indicates when the topic is given special consideration. Text figures are indicated by a page-number and asterisk: the figure numbers are not given. *Passim* signifies scattered throughout. Photographs are referred to at the end of each entry by their individual plate numbers: thus Pl. *21, 32* . . . English plant names (and some animal names) have usually been given their Latin equivalents in parentheses. Indexed Latin names, however, are not accompanied by their English equivalents. Both Latin and English name entries carry the same page references.

The positions of all the more important sites and waterways are shown in the text figures cited in the index; grid-references, however, are given for all the less familiar Fenland localities.

acidic mires, 13, 15–18, 20, 117, 182f, Pl. *8, 9, 31*
    see also peat, blanket bog, raised bog
adder, 171
Adelaide Bridge, Ely (TL 565 814), 60, Pl. *14*
Adventurers Fen, Wicken (TL 56 69), 60, 122f, 150, 163, 179, Pl. *39, 64*
ague, 157f
air-photography, 47, 88f, 98, 109
Alces alces (elk), 165f, Pl. *52*
alder *(Alnus glutinosa)*, 11, 14–39 *passim*, 56*, 103, 105, 128, 177, 182, Pl. *7, 24*
alder-buckthorn *(Frangula alnus)*, 16f, 151, 176, 183, Pl. *5, 51, 64*
alkalinity of fen peat, 15, 72
almond osier *(Salix triandra)*, 152
Andrew, Miss R., 57
Andromeda polifolia, 15, 169
angelica *(Angelica sylvestris)*, 16, 19, Pl. *10*
Anglo-Saxon period, 86, 89, 99, 135, 154
Anopheles maculipennis, 158
aquatic plants, 10, 11, 16f, 171f**, 183, Pl. *3, 7, 48*
arctic plants, 164
Artemesia, 57
artificial waterways, recognition of, 134
ash *(Fraxinus excelsior)*, 12, 18, 36, 39, 56*
Astbury, A.K., 87, 129, 134
Atlantic climatic period, 25, 50–9 *passim*, 104
aurochs *(Bos primigenius)* 60f*, 165, Pl. *17*
autogenic succession, 19
axes, prehistoric, 60f*
Azolla filiculoides, 9

Babington, C.C., 176
Barnack stone, 100
Barnwell Station beds (TL 470 596), 164
Barton Broad, Norfolk, 97, 116, Pl. *26*
Barway (TL 545 758), 75f, Pl. *20*
Bassenhally Fen (TF 290 005), 167
beaked sedges *(Rhyncospora* spp.), 15, 170
beam engines, 140
'bear's muck', 63
beaver *(Castor fiber)*, 166, Pl. *53*
becket, 115, 117–21**
Bedford, Earls of, 136
Bedford Levels, 81*, 85, 127, 129, 136*, 155, 184, Pl. *42*
Bedford Ouse, R., 137
beech *(Fagus sylvatica)*, 22*, 69, 98
bell-heather, 169
bench-marks, 132
Benwick (TL 34 90), 87, 129
Benwick Mere (TL 340 886), 71*, 91, 94
Bettisfield, 15*
Betula spp., 11, 16, 19, 22*, 36, 41 *et passim*, Pl. *11*
Bevill's Leam, 137
Bewick's swan, 184f
bilberry *(Vaccinium myrtillus)*, 169
birch *(Betula* spp.), 11, 16, 19, 22*, 36, 41, *et passim*, Pl. *11*
birch woods, 41, 93, 103, 105, 184
bison, 164
bittern, 167
black-tailed godwit, 184

black tern, 184
blackberry, 42
bladderwort, 150
blanket bog, 13, 111, 113
'blowing' of banks and sluices, 137, 139, 141
'blue buttery clay', *see* Fen Clay
Blytt and *Sernander* periods, 24f*
bog-asphodel *(Narthecium ossifragum)*, 170
bog-bean *(Menyanthes trifoliata)*, 172
bog-myrtle *(Myrica gale)*, 15, 71f, 169, 176, 183
bog-oaks, 29*, 33–42, 113, 127f, Pl. *14, 38*
Boreal climatic period, 25f, 32, 50–9 *passim*, 103
Borer, O., 47
Bos primigenius, 60f*, 165, Pl. *17*
Boston, 159, 162
Botrychium, 57
Boulder Clay, 37, 43, 58, 69, 102
brackish-water indicators, 51f, 62f, 105, Pl. *15*
Brandon, 100
Brandon Creek (TL 607 918), 86f, 95
Brassica napus var. *arvensis*, 153f
breast plough, 127
Breckland, 38, 57f, 66f, 94
Brick Mere (TL 233 860), 71*, 91, 93, 95*, 183
brick making, 122
brick pits, 13, 122, Pl. *48*
Bronze Age, 45, 68–78 *passim*, 105, 107, 114, 131*
    early, 50–9***, 65, 73, 106
    middle, 61*, 65*, 73f, 106
    late, 38, 61*, 73f*, 76f, 106, 156, Pl. *20, 21*
brown bear, 165
buckthorn, *see* alder-buckthorn and purging
    buckthorn
building stone, 100
bulrush *(Scirpus lacustris)*, 11, 96, 150f, Pl. *4, 48*
bur-reed *(Sparganium* spp.*)*, 10, Pl. *3*
buried forests, 29*f*, 33–42, 165
Burwell (TL 59, 67), 38, 60, 116f, 123, 166, Pl. *30*
Burwell Lode (TL 58 68), 87, 98, 163
Bush and tree colonisation, 11, 14, 17f, 181, 183,
    Pl. *5, 6, 62, 63, 64*
'Buttery Clay' *see* Fen Clay

Caistor ware, 85
calcareous marl, 30, 50*ff*, 70, 91–8, 130f*, Pl. *25*
Calluna vulgaris, 15, 71ff, 169, Pl. *9*
Calthorpe Broad, 19f, 27, 30, Pl. *10*
Cam, R., 58, 87, 99*, 137, 164
Camboritum (TL 707 885), 90
Cambridge Greensand, 38
Cambridgeshire and Isle of Ely Naturalists' Trust,
    184
canals, 85, 90, 98f*, 130, 134
Cannabis sativa, 154–7*, 171

Car Dyke, 87f
Cardium edule, 63, 65, 169
carnation sedge *(Carex panicea)*, 127, 181*
carrots, 132
Castle Hills Farm (TL 220 836), 61*, 74, 77
Castor fiber, 166, Pl. *53*
cat-tail, 11
cattle, 89f, 104, 115
causeways, 81*ff, 84*f
celery, 132
Cerambyx cerdo, 66, 165
Chalk, 69, 96, 99
Chara spp., 30, 91, 98
Chatteris (TL 395 860), 117
Chippenham Fen (TL 645 695), 40, 122
chronological scales, 23–5, 44*
Churchill, Dr D.M., 86
Cirsium palustre, 180
Clare, John, 96
Clark, Prof. J.G.D., 45f, 51, 53, 56, Pl. *16*
Cladium mariscus, 11, 14, 17, 112, 145*–9, 177,
    180, 182, Pl. *5, 28, 43–7, 51*
Clayhithe (TL 502 644), 87
Clifford, Dr M.C.H., 63f
climatic indicators, 21, 23f, 66f, 165
climatic periods, 25, 44*
    see also individual entries
climax vegetation, 12
Cnut's Drain (TL 250 920), 100
coal, 111, 116
cockle *(Cardium edule)*, 63, 65, 169
codein, 156
cole (rape) *(Brassica arvensis* var. *napus)*, 153f
collectors, 17, 177f, 180
colza-oil, 154
coprolites, 38, 60
corn, 87, 90
Cottenham (TL 455 675), 88, 152
cotton-grass *(Eriophorum* spp.*)*, 15, 70ff, 127,
    169f, 176, Pl. *8, 18*
county boundaries, 95
Coveney (TL 490 823), 119
cowberry *(Vaccinium vitis-idaea)*, 169
cranberry *(Vaccinium oxycoccus)*, 15, 96f, 168ff,
    Pl. *55, 56*
crane, 167
Crawford, O.G.S., 47
creek patterns, 64*, 88, 92, 109f, Pl. *23, 25*
creeping willow *(Salix repens)*, 16, 175
cropping of vegetation, 18, 178f, 181*, Pl. *62*
Cross Bank, Shippea Hill (TL 660 836), 93
Crosswater Staunch (TL 676, 856), 95
cross-leaved heath *(Erica tetralix)*, 15, 169, Pl. **9**
crowberry *(Empetrum nigrum)*, 5, 169

Crowland Abbey (TF 242 103), 3, 108, 115
Cut off Channel, 50*, 81*, 136*, 142f, Pl. *42*

Darby, Prof. H.C., 47, 114, 135f*, 160f
'*Darg*', 63
deflected succession, 180*
Denver (TF 615 017), 65, 81*, 84f
Denver Sluice (TF 590, 010), 137, 139, 142, Pl. *42*
Deschampsia caespitosa, 127
Devensian period, 164
devil's-bit scabious *(Succisa pratensis)*, 180f*
diatoms, 63f, 167
Dogger Bank, 25f*, 57, 103, 109
dogwood *(Cornus sanguinea)*, 42
Domesday Survey, 91, 94f, 135, 160
Doran, W.E., 47, 140
drainage, 101, 107f, 124, 132f, **134–44***
    'Commissioners' Drains', 141
    consequences, **127–33**, 137–41, 169, Pl. *36, 37*
    'Engine Drains', 141
    perched water-ways, 141
    pumps, 138ff, Pl. *38–41*
Drayton, M., 155, 161
Drosera rotundifolia, 15, 169ff, 176, Pl. *54*
drug-plants, 155–8**
dry-land phase of N. Sea, 25, 26*, 164
duck decoys, 162, 814
duck-gun, 162
Duffey, Dr E.A., 171
'dydal', 122
dwarf birch *(Betula nana)*, 25, 57
dwarf willow *(Salix repens)*, 16, 175

Earith (TL 390 748), 87, 136f, 164
East Anglian Water Authority, 143f
East Fen, S. Lincs (TF 41 56), 96, 114
ecclesiastical houses, 100, 108, 115, 135, 159, 161
ecology, 4, 6, **9–20**, 45, 177–86
Edwards, W. (Will'En), 33, 77f, 117f, 122, 139
eels, 161
eel-spear (glaive), 161
elk *(Alces alces)*, 165f, Pl. *52*
elm *(Ulmus* spp.*)*, 22*, 103f, Pl. *11*
Ely, 2f, 75f, Pl. *1*
Empetrum nigrum, 169
Emys orbicularis, 165
Enclosure Acts, 115
Ennion, Dr E.A., 117, 123, 163, 166
Ephedra spp., 57
Erdtman, G.E., 21
Erica cinerea, 169
Erica tetralix, 15, 169, Pl. *9*
ericoid plants, 15, 68f

Eriophorum angustifolium, 15, 127, 170, 176, Pl. *9*, *19*
Eriophorum vaginatum, 15, 170, Pl. *9*, *18*
estuary of the Fenland, 2*, 81f, 90, 107, 134
Eupatorium cannabinum, 148
eustatic change in sea-level, 25ff, 57ff, 104ff*, 109f
extinct waterways, 86
    see also old runs, old slades, roddons

Fagus sylvatica, 22*, 69, 98
Farren, W.S., 178
Feltwell Fen (TL 650 900), 66f
fen-birch *(Betula pubescens)*, 39, 41
'fen-blows', 133
fen carr, 12, 17–20, 28*, 32, 105, 181, Pl. *6, 7, 62*
Fen Clay, 5, 50–9 *passim*, **60–7**, 68–75 *passim*, 92f, 103*, 105f, 108ff, 131, 144, 167, 169, Pl. *25*
fen-margin economy, 152f
fen oakwoods, 12f, 18ff, 27ff**, 30f, Pl. *10*
'fen slodger', 161, 163
fen violet *(Viola stagnina)*, 172
fen woods, 12f, 18ff, 27f*f*, 30, 38–42, 105, 181
Fenland Research Committee, 7, 34, **45–8**, 55, 60, 62, 68, 73, 82, 88, 108, 126, 184
field patterns, 82*, 88f, Pl. *24*
fishing, 91, 94f, 100, 160f
fish weirs, 161
fisheries, 91, 94f, 100, 161
Flagrass (TL 434 985), 81*, 84
Flandrian Period, 22*, 24*, 43ff**, 102–10**, 165
flax *(Linum usitatissimum)*, 155f
flooding, 142, 181, 184
foraminifera, 27, 51f, 62f, 85, 167, 169, Pl. *15*
forest horizons, 29*, 30*
Forty foot Drain, 137
Fowler, Maj. Gordon, 29, 35, 45f, 50, 53, 79–83, 86, 93, 95, 98, 124, 130, 134, Pl. *16*
Fox, Sir Cyril, 45, 73, 79
Frangula alnus, 16f, 151, 176, 183, Pl. *5, 51, 64*
Fraxinus excelsior, 12, 18, 36, 39, 56*
fringed water-lily *(Limnanthemum peltatum)*, 172, 185
Friskney (TF 460 555), 97, 114
frogbit *(Hydrocharis morsus-ranae)*, 9, 172*, 185
frost wedges, 164

'gatways', 16, 94*, Pl. *7*
Gault Clay, 102
'*Gentlemen Adventurers*', 137
geology, 12*–16 *et passim*
German coastal deposits, 6*, 83
giant Irish deer, 164f
giant reed *(Phragmites communis)*, 11, 28, 30, 63, 96, 105, Pl. *2, 10*

giant sedge, *see* sword sedge
Glacial Period, 38, 102
Glass Moor (TL 295 930), 40, 62f, 169
Glastonbury, 159, 167
'glastrum', *see* woad
Glen, R., S. Lincs., 89
goose keeping, 162
grampus, 63
Granta, R., 86f
great fen ragwort *(Senecio paludosus)*, 172f, Pl. *57*
Great Ouse, R., 81*, 86f, 91, 128f, 136, Pl. *36, 37, 42*
great water-dock *(Rumex hydrolapathum)*, 175
greater spearwort *(Ranunculus lingua)*, 175
Green Dyke, Lotting Fen (TL 245 863), 64*, 93, Pl. *15*
Greensand, 38, 102
Grunty Fen (TL 510 800), 75
guelder rose *(Viburnus opulus)*, 16f, Pl. *47*
'gurgites' *see* fish weirs
Guyhirne (TF 400 035), 64, 135

hair-moss *(Polytrichum* sp.), 20
hairy birch *(Betula pubescens)*, 11
Hallam, Dr S.J., 88f
Harrimere (TL 535 745), 91
hassock plough, 127
hazel *(Corylus avellana)*, 22*, 36f, 42, 103, Pl. *11*
heather (Erica spp., *Calluna vulgaris)*, 63, 169
Helianthemum, 57
hemp *(Cannabis sativa)*, 154-7*, 171
hemp-agrimony *(Eupatorium cannabinum)*, 16, 148*
'heneplond', 155
Heron's Carr, Norfolk, 14, Pl. *6*
hides, 90
high forest, 35ff, 60ff, 104, 165f, 182, Pl. *14*
Hippuris vulgaris, 176, Pl. *61*
hoards, 61, 74f
Hockham Mere, 67, 97
Hockwold (TL 707, 885), 94
Hockwold-cum-Wilton (TL 735 880), 50*, 90
Holbeach, 157
Holbeach St. John's (TF 34 17), 89, Pl. *24*
Holme Fen (TL 205 890), 72f, 105, 107, 112f, 122f, 130f*, 169, Pl. *25, 27*
Holme Fen post (TL 203 894), 124, 131*, 183f, Pl. *35*
hornbeam *(Carpinus betulus)*, 22*, 69, 98
Hoveton Broad, Norfolk, 16, 116, Pl. *7*
hulled barley, 89
humification, 15*, 113
'Hundeköd', 63
'Hundred Foot River', 136

Hunts. and Beds. Naturalists' Trust, 184
Hydrocharis morsus-ranae, 9, 172*, 185
Hydrocotyle vulgaris, 148*, 180f*

indigo *(Indigofera tinctoria)*, 160
Iron Age, 79, 86, 107ff, 144, 159, 167
Isatis tinctoria, 158ff
Isleham Fen (TL 630 765), 38, 60, 66, 119

Jackson's Fen, Woodwalton (TL 23 84), 120
Jennings, Prof. J.N., 56, 94, 96
Jones, Prof. O.T., 24, 46
Juncus obtusiflorus, 181*
Jurassic Clays, 37, 102, 124

Kenny, Dr E.J.A., 82*
Kimmeridge Clay, 37
King's Delph (TL 240 954), 100

'lagg-stream', 177
lagoon conditions, 52, 62, 105
lake-muds, 11f
Lakenheath, 94
Lakenheath Lode (TL 847 680), 93
Landbeach (TL 477 653), 152
landscape quality, 2ff, Pl. *1, 2*
Large Copper butterfly *(Lycaena dispar)*, 175
Lark, R., 86, 137, 142
'Late-glacial' period, 26*, 57f, 102, 164ff
vegetation of, 57, 102
laudanum, 156
Lathyrus palustris, 175, Pl. *60*
'laving' ('lecking'), 121, Pl. *34*
leaf-fodder, 104
'leck', 121f, Pl. *34*
Leman and Ower bank, 26*
lemming, 164
Lethbridge, T.C., 46
levées, *see* roddons
lignite, 111
Lime *(Tilia* spp.), 22*, 24*, 36f, 56* *et passim*, **66**, 69, 104, 107
Limnanthemum peltatum, 172, 185
Lincoln, 87
linden, *see* Lime
ling *(Calluna vulgaris)*, 15, 68, 71f, 169, Pl. *9*
Linum usitatissimum, 156f
Liparis weselii, 175f
'litter' community, 181*f
Little Ouse, R., 50-9***, 60, 82, 86f, 91, 93ff, 103, 107, 137, 142, Pl. *16, 22, 23*
Little Thetford (TL 532 763), 75f, Pl. *20*
Littleport (TL 566 870), 117, 128f

Lord Orford's voyage, 100f
lost and vanishing species, **164–77**
Lotting Fen (TL 245 853), 33, 93, 122
Lower Peat, 50–9* *passim*, **60–7**, 93, 102–5*
'loy' *see* becket
Lycaena dispar, 175
Lysimachia vulgaris, 16, 175, Pl. 60
Lucas, Dr C., 130
Lycopodium sp., 57
Lycosa paludicola, 171
Lynn Ouse, R., 86, 95, 142, Pl. 42

'mabs', 127, Pl. *18*
Macfadyen, Dr W.A., 27, 46, 167
macrofossil plant remains, 27, 29, 39–42, 93, 113
malaria, 157f
mammoth, 164
management of reserves, **178–85**
March, 65, 84
mare's tail (*Hippuris vulgaris*), 176, Pl. *61*
Marginal Relief Channel, 81*, 136*, **143**, 164,
    167, Pl. *42*
marine deposits in Fenland, 1f*, 27* *et passim*
marine transgression, 5f, 25f, 32, 38, 41, **60–68**,
    77, 79–90, 103–10** *et passim*
marine regression, 5f, 32, 73, 105ff*
market garden crops, 160
marsh fleawort (*Senecio palustris*), 172
marsh pea (*Lathyrus palustris*), 175, Pl. 60
marsh pennywort (*Hydrocotyle vulgaris*), 148*,
    180f*
marsh sowthistle (*Sonchus palustris*), 172
Marshall, W., 29, 71f, 169
Marshland, 5, **87–90**, 108, 129, 137*, 149, 155,
    160, 169, Pl. *24*
meadow thistle (*Cirsium palustre*), 180
medieval artefacts, 108, 135
Menyanthes trifoliata, 172
meres, 71*, **91–101***, 108
Mesolithic period and cultures, 26, 43, 45, 51–9*
    passim, 103, 166, Pl. *16*
Methwold (TL 733 948), 73, 94f, 137
microliths, 51, 103
Middle Ages, 88, 94f, 107f, 114, 122, 135, 144,
    155, 158, 161, 165
migratory birds, 184
Mildenhall Fen (TL 665 780), 60, 74, 165
milk parsley (*Peucedanum palustre*), 174f, Pl. *58,
    59*
Miller, S.H. and Skertchly, S.B.J., 3, 29, 161
mires, 13, 111
'mixed-sedge' communities, 180*, Pl. *5*
Moberley, J.C., 178
Molinia caerulea, 127, 148*, 180f*

mollusca as environmental indices, 55, 63, 65, 91,
    169
Monks' Lode, Sawtry (TL 212 858), 95*, 135
Monk's Lode, Wicken (TL 565 700), 115, 135
moorlog, 25f*, 58
morphine, 156
Morton, Bishop, 135
Morton's Leam (TL 250 984), 135
mosquito as vector of malaria, 158
moth-collecting, 177–80, Pl. *62*
Myrica gale, 15, 71f, 169, 176, 183

Naias marina, 67
'Nancy', 65*, 68, 73
Narthecium ossifragum, 170
National Trust, 151
natural crops, **145–53**
Nature Conservancy, 183f
Nature Conservancy Council, 184
nature reserves, **178–85**
Nene, R., 82, 87, 91f, 94f, Pl. *25*
Neolithic Period and culture, 35, 37f, 45, 52–9**
    passim, 60f**, 104
New Bedford River, 136*f, 169, Pl. *42*
New South Eau (TF 310 086), 137
Nordelph (TF 558 010), 86f*, 105
Norfolk Broads, 94, 96f, 114, 150, Pl. *10, 26*
'Norfolk reed', 146, 150
North Fen, *see* Wood Fen
North Sea, 25f*, 57, 102
Nuphar luteum, 10f, 176, Pl. *3, 61*
Nymphaea alba, 10f, 150, Pl. *3, 4, 48*

oak (*Quercus* sp.), 12f, 18ff, 22*, 27*f*f*, 103, 105
    et passim, Pl. *10, 11*
oil crops, 153–6, 158
Old Bedford River, 136*f, Pl. *42*
Old Croft River, 81
Old Decoy, Cambs. (TL 665 855), 56*f
Old Nene River, 71*, 91f, 94f*, 134, Pl. *25*
'old runs', 68
'old slade', 87, 98, 130
Old West River (TL 500 710), 87
oligotrophic mires, 15
ombrotrophic mires, 15
Opium poppy (*Papaver somniferum*), 156–8*
Orobanche ramosa, 171
osiers, 152
otter, 166f
Ouse Wash Reserves, 184f
Outwell (TF 513 036)
'overkill', 166
Oxford Clay, 37
oxygen deficiency, 111, 125f, 146

palaeoecology, 6
Papaver somniferum, 156ff*
Papilio machaon, 173ff, Pl. *58, 59*
paring and burning of peat, 127
paring plough, 127, 154
Parish boundaries, 95*, 97, 124f
Parsonage Drove, Wisbech (TF 390 090), 160
Peacock's Farm, Shippea Hill (TL 628 847), 51,
    56*, 65f, 73
Peakirk Drain (TF 280 050), 137
peat, 111–23 *et passim*
    acidification of, 12*, 15, 20, 30*f, 40, 68–73**,
      93, 113, 123, 169, 181ff, Pl. *18, 19*
    'black' and 'white', 15*, 113
    'cesses', *see* 'turves'
    cutting tools, 117f**
    differential contraction, 82, 98, 128–33*, 141,
      Pl. *36, 37*
    digging, 17, 96ff, 111, 114–23**, 169, 176f, Pl.
      8, 26, 29–32
    drying of cut, 115, 121f, Pl. *33, 34*
    fen, 12*, 15, 111, 113 *passim*
    fires, 35, 127
    'hoddy', 118* ff
    loss, 124–33
    origin of, 5, 11f*, 58, 111
    'shrinkage' of, 82f, 124–33*, 139
    '*Sphagnum*', 12*, 15, 30*, 40, 70ff, 93, 96, 113,
      165, Pl. *18, 19*
    'turves', 114f, 118*–22, Pl. *29, 30*
    types of, 12*, 70, 112f, Pl. *18, 27, 28*
    wastage of, 63, 69, 78, 83, 90, 92, 94, 98, 109,
      124–33, 139, 141, Pl. *36, 37*
    water-transport of, 120
pelican (*Pelicanus* sp.), 66, 167
Pencedanum palustre, 174f, Pl. *58, 59*
Peterborough, 81f, 87, 99, 135
Phillips, C.W., 46, 85
Phragmites communis, 11, 14, 28, 30, 63, 96, 105,
    146, 148*–50, 162, 181*, Pl. *2, 10, 48–50*
pine (*Pinus sylvestris*), 22, 29*ff**, 40f, 62, 69*,
    77f, 105, Pl. *12*
pine (–birch) woods, 30*, 32, 40f, 78, 93, 103
'piper', 162
Plantation Farm, Shippea Hill (TL 642 847), 50*,
    65, 73, Pl. *16*
Plav, 149
polished stone axes, 60f**, 104, Pl. *17*
pollen analysis, 6, 21–32*** *et passim*, Pl. *11*
    local factors affecting, 27ff**, 32, 63
    principles, 21
    zonation, 23, 25*, 56*, 58*
pond tortoise (*Emys orbicularis*), 66f, 165
pondweeds (*Potamogeton* spp.), 185

Polytrichum spp., 20
'Poor's Pieces', Wicken, 115, 146, 148, 176
Popham's Eau, 68, 136*
Poplar Farm, March (TL 442 959), 81*, 83f*f
'post-glacial', *see* Flandrian
'poy-stick', 162
Pre-boreal period, 25, 57
Prickwillow (TL 597 825), 86f, 100
primary succession, 13, 17ff, 179f*
'primeval fenland', 177, 183
Pseudamnicola confusa, 167
Pseudorca crassidens, 66
'pure sedge' community, 17
purging buckthorn (*Rhamnus cathartica*), 17, 42,
    151f
purple loosestrife (*Lythrum salicaria*), 16
purple moor-grass (*Molinia caerulea*), 127, 148*,
    180f*
purple osier (*Salix purpurea*), 152

Quaternary Epoch, 45
Quaternary research, 44
quays and staithes, 87, 99
Queen Adelaide Bridge, Ely (TL 565 814), 60, Pl.
    14
Queen's Ground, Methwold (TL 689 930), 73
Quercus, 12f, 18ff, 22*, 27*ff*, 103, 105 *et passim*,
    Pl. *10, 11*
Quy (TL 516 599), 162

radiocarbon dating, 25f, 36, 40, 44*, 48, 52*, 56*,
    60ff, 65f, 76, 78, 85f, 93, 103*, 108, 165
    age-citation, 48f
raised bog, 13, 15*, 20, 30, 32, 68–72, 77, 93, 96f,
    105, 113, 123, 131*, 169ff, 183, Pl. *8, 9, 27,
    31, 32, 33*
Ramsey Heights (TL 245 847), 77, 122
Ramsey Mere (TL 310 890), 71*, 91, 94f, 108
Ramsey St Mary's (TL 256 880), 130
Ranunculus lingua, 175
rape (*Brassica napus* var. *arvensis*), 153f
Raveley Drain, Woodwalton (TL 234 840), 120
Ravensdale, J.R., 152
Ray, John, 172, 176
Reach Fen (TL 558 680), 69, 87, 123, 130
Reach Lode (TL 555 680), 87, 98f*, 163
'reaction' in vegetational succession, 10–12, 18, 32
red deer, 165
Red Mere (TL 670 840), 50*, 52*, 84*, 91, 94*f,
    130
reed (*Phragmites communis*), 11, 28, 30, 63, 96,
    105, 146, 148*–50, Pl. *2, 10, 48–50*
reed-mace (*Typha* spp.), 11
reed-swamp, 16f, 19, 96, 149f, Pl. *7*

reed-warbler, 150

reindeer, 164

Relief Channel, Marginal, 81*, 136*, 143, 164, 167, Pl. *42*

Rhamnus cathartica, 17, 42, 151f

Rhododenron, 20

Rhyncospora alba, 15, 170

roddons (rodhams), 50*–59 *passim*, 69*, 79–90, 94ff, 110, 126, 129f, 169, Pl. *22, 23*

    origins of, 82ff*

Rodham Farm, March (TL 460 981), 81*, 84, 94

roebuck, 165

Roman bridge, Nordelph (TF 573 004), 69*, 82f*

Roman causeway, Denver–Peterborough, 81–5**

Romano-British Period, 52*, 55, 68, 79–90, 107ff, 128, 130, 134f, 144, 182, Pl. *36, 37*

rose, 42

Rothschild, the Hon. Charles, 183

Royal Society for the Protection of Birds, 184

Rumex spp., 57

Rumex hydrolapathum, 175

Saddlebow (TF 408 157), 65f

St Guthlac, 3, 108

St German's (TF 590 148), 26ff**, 62f, 65, 68, 88

Salix cinerea, 11, 16f, 19, 29, 39, 104, 152, Pl. *5, 7*

Salix purpurea, 152

Salix repens, 16, 175

Salix triandra, 152

Salix viminalis, 152

sallow *(Salix cinerea)*, 11, 16f, 19, 29, 39, 104, 152, Pl. *5, 7*

salt production, 90, 114, 122

Salter's Lode (TF 585 017), 143, Pl. *42*

salt-marsh, 28, 64*, 88f, 93, 107f, Pl. *23, 24*

Salway, P., 89f

Sam's Cut, 137

Sawtry (TL 167 837), 95*, 100

Scheuchzeria palustris, 112, 170, Pl. *27*

Scrobicularia piperita, 169

sea-level changes, 1, 5, 6*, 24–8, 38, 52 *et passim*, 106*, 144

sedge communities, 104, 146, 148*, Pl. *47*

sedge cropping, 115f, 145–9*, Pl. *44, 47*

sedge-warbler, Pl. *65*

seed viability, 172f

Senecio paludosus, 172f, Pl. *57*

Senecio palustris, 172

Seward, Sir A.C., 34, 45, Pl. *13*

sewers (drains), 135

Shapwick Heath, Somerset, 120f, Pl. *33, 34*

sheep, 89

shell-marl, 50*ff*, 70, 91–8, 130f*, 165, 182 Pl. *25*

Shippea Hill, Cambs. (TL 640 840), 26, 47, 50–59***, 60, 65, 73, 80, 84*, 93, 103f, Pl. *16, 22, 23*

Silt land, 2*, 5, 87–90, 108, 129, 137, 144, 155, 160, 169, Pl. *24*

Sixteen foot Drain, 137

skating, 101

Skertchly, S.B.J., 3, 29, 45, 50, 92, 117, 119, 122

'slane', *see* becket

Society for Promotion of Nature Reserves, 183

Soham Mere (TL 594 732), 91, 96

Soil Survey maps, 64, 110

Sonchus palustris, 172

South Level, 142f

Southery (TL 622 946), 50*, 65*, 73

Spalding, 156, 159

Sparganium simplex, 10

spelt wheat, 89

Sphagnum moss, 15, 20, 30f, 40, 96, 113f, 169, Pl. *9, 55, 56*

Steers, Prof. J.A., 47

stone axes, 38

stratigraphy, 45, 48 *et passim*, 103*

Stratiotes aloides, 171f*

Streatham Mere (TL 523 727), 91, 108, 130

Stowbridge (TF 602 075), 86f

Stuntney (TL 555 780), 61*, 74, Pl. *1*

submerged forests, 27*f*

submerged peat-beds, 24–8* *passim*

subsidence of banks, 129, 132f, 141

    of buildings, 128f, Pl. *36, 37*

    of railways, 131f

Sub-atlantic period, 25, 32, 107f

Sub-boreal period, 25, 31, 68–78 *passim*, 105–7

succession, *see* vegetational succession

Succisa pratensis, 180f*

summer lime *(Tilia platyphyllos)*, 22*, 26

sundew *(Drosera* spp.), 15, 169–71, 176, Pl. *54*

Swaffham Engine drain (TL 540 695), 61, Pl. *41*

Swaffham Fen (TL 540 673), 115, 123, 162, Pl. *29*

swales, 36

swans, 185

Swansea Bay, 24f, 58

swallow-tail butterfly *(Papilio machaon)*, 175ff, Pl. *58, 59*

sweet gale *(Myrica gale)*, 15, 71f, 169, 176, 183

Switsur, R.A., 48

sword sedge *(Cladium mariscus)*, 11, 14, 17, 112, 145–9, 177, 180, 182, Pl. *5, 28, 43–7, 51*

Tansley, Sir Arthur, 9

Taxus baccata, 29, 33, 38ff, 73, 105, 166, Pl. *13*

Terrington St Clements (TF 552 205), 129

Thalictrum sp., 57

Thames estuary, 40, 144
'thermal maximum', **66–7**
thermophilous organisms, **66–7**, 165
Three Holes Farm (TL 522 974), 82
tidal action, 52, 63, 82–90 *et passim*, 94, 105, 107, 137, 139, 142–4, 169
tidal model of Wash, 143
tidal surges, 142
Tilia cordata, **22***, 66
Tilia platyphyllos, **22***, 66
tilting of houses, 128f, Pl. *37*
topogenous mires, 15
trackways, 75ff, 106, Pl. *20, 21*
'treacling' for moths, 178
tree-pollen identification, **22***, Pl. *11*
tree-ring analysis, 39f
Tregaron Bog, Cardigan, 17f, Pl. *8, 9*
Trichophilous paludosum, 171
Trundle Mere (TL 205 915), 71***, 91, 95*f, 123, 169
tufted hair-grass *(Deschampsia caespitosa)*, 127
turbaries, 114
turbary rights, 90, 114f
'turf', *see* peat
turf-knife, 118****f
turf-spade, 118****f, 121, Pl. *29*
turves (cesses), 114f, 118*–22, Pl. *29, 30*
tussock sedge *(Carex paniculata)*, 127
Twenty foot River, 137
Typha angustifolia, 11

Ugg Mere (TL 245 870), 40, 63f***, 70f***, 78, 91, 95***, 165, Pl. *15*
Upper Peat, 50–9** *passim*, 65, 68–78, 93, 103***, 105f
Upper Silts, 108f
Upware (TL 537 700), 9, 61, 87, 98, 100
Upware Lode (TL 540 697), 87, 163
'Upware Republic', 163
Upwell (TF 505 027), 81

Vaccinium oxycoccus, 15, 96, 168–70, Pl. *55, 56*
Vaccinium myrtillus, 15, 169
Vaccinium vitis-idaea, 169
'varves', 44
vegetational succession, **10–20***, 28*, 30ff, 38, 45, 92, 104, 111, 179
Vermuyden, Cornelius, 124
Verrall, G.H., 178
Viburnum opulus, 16f, Pl. *47*
Viking period, 43
Vine, A., 93
Viola stagnina, 172
viper (adder) *Viperus verus*, 171

'wad', *see* woad
Wadham, Sir Samuel, 179
Wainfleet, 96, 114
Walpole St Peter (TF 502 169), 129
Wash, The, 107, 109, 143f
'washes', 101, 137, 161, 184
wastage of peat, *see* peat
water-fowl, 160–3, 167, 179, 184
water-milfoil *(Myriophyllum* spp.), 148
water-soldier *(Stratiotes aloides)*, 171f*
water transport, 83, 85ff, 92, 100, 107, 116, 135, 148, 182, Pl. *44*
Waterbeach (TL 498 651), 87, 152
weeds, 106, 171
Welland, R., 87
Wells, S., 71f, 169
Wells, W., 91, 101, 124, 131
Welche's Dam (TL 470 859), 137
Welney (TL 526 940), 81f, 85f, 94, 107, 184
Welney Washes (TL 531 935), 85f
West Fen, S. Lincs. (TF 31 54), 96
Westhay, Somerset, 77, Pl. *21*
whales, 63, 105, 167*
white water-lily *(Nymphaea alba)*, 10f, 150, Pl. *3, 4, 48*
Whittlesey (TL 270 970), 70f*, 82, 99f, 155
Whittlesey Mere (TL 225 905), 63, 70f*, 82, 91–100 *passim*, 108, 124f, 130f*, 169, 171, 183, Pl. *25*
Whixall raised bog, 117, Pl. *31*
whortleberry *(Vaccinium myrtillus)*, 15
Wicken Fen (TL 553 704), 6, 9, 13f, 16ff, 39, 69, 98, 115, 120, 122, 146–52*, 163, 171–83, Pl. *3, 39, 40, 44, 46, 47, 61–4*
Wicken Lode, 10, 98, 146, 151, 176, 182, Pl. *3, 44, 49, 61, 64*
wild boar, 165
wild horse, 164
wildfowl as crop, 160–3
Wildfowl Trust, 184
William the Conqueror, 75
Willingham Mere (TL 403 733), 91
Willis, E.H., 48
willows, 29, 36, 105, 182
  see also osiers, sallow
Wilton Bridge, Cambs (TL 724 867), 55, 63, 69, 98
wind-pumps (windmills), 138f, Pl. *38–40*
winter buds of aquatics, 171f*
winter lime *(Tilia cordata)*, **22***, 66
Wisbech, 64, 81, 107, 135, 154, 157, 159
Wisbech Ouse, R., 86
Wissey, R., 137, 142
Witham, R., 87
woad *(Isatis tinctoria)*, 185ff

wolf, 165

wooden tub, Bronze Age, 75*

Wood Fen (North Fen) (TL 550 850), 29*f*f*, 35, 39f, 50*, 62, 65, 124f, 169, Pl. *12*

wood-peat, 39

woodland clearance, 44*, 78, 106, 113, 131*

Woodwalton Fen (TL 230 850), 40, 70–8 *et passim*, 93, 120, 122f, 169, 172, 175f, 183

wool, 90, 159f

woolly rhinoceros, 164

Wrangle, S. Lincs. (TF 425 509), 114

Wretham Mere, 66f

Wretton (TF 690 000), 164

Wroxham Broad, Norfolk, 16, Pl. *7*

Yapp, R.H., 10, 13

yellow loosestrife *(Lysimachia vulgaris)*, 16, 175, Pl. *60*

yellow water-lily *Nuphar luteum)*, 10, 176, Pl. *61*

yew *(Taxus baccata)*, 29, 33, 38ff, 73, 105, 166, Pl. *13*

zonation of vegetation, 13, 16, 19, Pl. *5–7*

Zwischenmoorwald, 30, 32